Advances in Desalination Insights

Edited by Huijin Xu, Xiaokun Gu, Qing Li, Hui Wang and Yasong Sun

Published in London, United Kingdom

Advances in Desalination Insights
http://dx.doi.org/10.5772/10.5772/intechopen.1002090
Edited by Huijin Xu, Xiaokun Gu, Qing Li, Hui Wang and Yasong Sun

Contributors
Amira S. Mahmoud, Amr Seoudy, Ashraf Morsy, Bin Lu, Haixiang Feng, Hamida Abdel-Hamid, Hao Luo, Huijin Xu, Hui Wang, Jiajiong Xu, Jianbo Lu, Mahmoud Shaban, Qing Li, Savan K. Raj, Shaoxuan Xu, Vishwakarma Ravikumar Ramlal, Xiaokun Gu, Xiao Luo, Yunqi Li, Yu Qiu, Zhiguo Qu

First published in London, United Kingdom, 2025 by IntechOpen
IntechOpen is the global imprint of INTECHOPEN LIMITED, registered in England and Wales, registration number: 11086078, 167-169 Great Portland Street, London, W1W 5PF, United Kingdom

For EU product safety concerns: IN TECH d.o.o., Prolaz Marije Krucifikse Kozulić 3, 51000 Rijeka, Croatia, info@intechopen.com or visit our website at intechopen.com.

British Library Cataloguing-in-Publication Data
A catalogue record for this book is available from the British Library

Advances in Desalination Insights
Edited by Huijin Xu, Xiaokun Gu, Qing Li, Hui Wang and Yasong Sun
p. cm.

This title is part of the Environmental Sciences Book Series, Volume 26
Topic: Water Science
Series Editor: J. Kevin Summers
Topic Editor: Yizi Shang

Print ISBN 978-0-85014-727-8
Online ISBN 978-0-85014-726-1
eBook (PDF) ISBN 978-0-85014-728-5
ISSN 2754-6713

If disposing of this product, please recycle the paper responsibly.

IntechOpen Book Series

Environmental Sciences

Volume 26

Aims and Scope of the Series

Scientists have long researched to understand the environment and man's place in it. The search for this knowledge grows in importance as rapid increases in population and economic development intensify humans' stresses on ecosystems. Fortunately, rapid increases in multiple scientific areas are advancing our understanding of environmental sciences. Breakthroughs in computing, molecular biology, ecology, and sustainability science are enhancing our ability to utilize environmental sciences to address real-world problems.

The four topics of this book series - Pollution; Environmental Resilience and Management; Ecosystems and Biodiversity; and Water Science - will address important areas of advancement in the environmental sciences. They will represent an excellent initial grouping of published works on these critical topics.

Meet the Series Editor

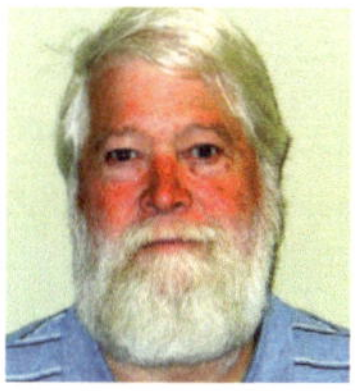

J. Kevin Summers is a Senior Research Ecologist at the Environmental Protection Agency's (EPA) Gulf Ecosystem Measurement and Modeling Division. He is currently working with colleagues in the Sustainable and Healthy Communities Program to develop an index of community resilience to natural hazards, an index of human well-being that can be linked to changes in the ecosystem, social and economic services, and a community sustainability tool for communities with populations under 40,000. He leads research efforts for indicator and indices development. Dr. Summers is a systems ecologist and began his career at the EPA in 1989 and has worked in various programs and capacities. This includes leading the National Coastal Assessment in collaboration with the Office of Water which culminated in the award-winning National Coastal Condition Report series (four volumes between 2001 and 2012), and which integrates water quality, sediment quality, habitat, and biological data to assess the ecosystem condition of the United States estuaries. He was acting National Program Director for Ecology for the EPA between 2004 and 2006. He has authored approximately 150 peer-reviewed journal articles, book chapters, and reports and has received many awards for technical accomplishments from the EPA and from outside of the agency. Dr. Summers holds a BA in Zoology and Psychology, an MA in Ecology, and Ph.D. in Systems Ecology/Biology.

Meet the Volume Editors

Dr. Huijin Xu is a professor at Shanghai Maritime University in China. He graduated from Xi'an Jiaotong University. He has worked at China University of Petroleum and Shanghai Jiao Tong University. His research interests cover Marine Hydrogen and Energy Storage, Shipborne Carbon Capture, Porous Media, etc. He significantly contributed to heat and mass transfer in porous media and thermal storage. He has published nearly 100 papers in high-level international journals, receiving more than 4000 citations (Google Scholar). He now serves as AE for two international journals: Journal of Porous Media and Special Topics and Reviews in Porous Media.

Dr. Xiaokun Gu is a Professor at the School of Energy and Materials at Shanghai Polytechnic University in China. Dr. Gu received his Ph.D. degree from the University of Colorado at Boulder in 2016. His research focuses on nanoscale energy transport and thermal management fundamentals, with particular expertise in phonon thermal conduction and solar thermal utilization. Dr. Gu has led two projects funded by the National Natural Science Foundation of China. He has authored over 60 SCI-indexed publications, with over 30 as first or corresponding author. According to Google Scholar, his work has garnered significant attention in the scientific community, with over 5,600 citations and an H-index of over 30.

Prof. Qing Li is a full professor at the School of Energy Science and Engineering at Central South University. He obtained his Ph. D degree from Xi'an Jiaotong University. Subsequently, He worked at the University of Southampton as a Postdoctoral Research Fellow from May 2011 to May 2014. After that, he joined Central South University. He mainly specializes in heat transfer enhancement and solar energy utilization. He was awarded the National Science Fund for Excellent Young Scholars by the National Natural Science Foundation of China in 2018. He has published over 130 internationally refereed journal papers, including more than 100 as the first or the corresponding author. He also holds 21 patents and co-authored two books.

Dr. Hui Wang is a distinguished researcher from the Xi 'an Jiaotong University in China. His research focuses on heat and mass transfer mechanisms and intelligent regulation at complex interfaces. As the first/corresponding author, he has published 60 papers in SCI index journals such as Nature Communication and Advanced Functional Materials, 2 ESI high award papers, and 2 journal excellent paper awards. He was awarded the second-class general of the China Engineering Thermophysics Society (ranked 2nd) and the second prize in the Science and Technology Progress Award of China Aeronautical Society (ranked 2nd). He served as an expert in preparing the outline of the roadmap of hydrogen energy technology development, the youth editor, and the special issue editor of many journals.

Dr. Yasong Sun is a professor at Chang'an University. He is a visiting scholar at Rutgers University, Oxford University and Nottingham University. He is a Shaanxi Youth Science and Technology Workers Association director at the Shaanxi Society of Automotive Engineering. His research interests are thermal management, rapid cooling, and computational fluid dynamics in energy and power systems. He has published over 80 journal papers. Over 60 of these papers have been published in SCI journals. These published papers have been cited over 1000 times in the SCI database. He has received the following academic honors: Shaanxi Higher Education Outstanding Young Talent, International Clean Energy Talent, Excellent Young Teacher Award of Henry Ying Tung Fok Education Foundation, Young Chang'an Scholar.

Contents

Preface

In an era where water scarcity is an increasingly pressing global concern, the pursuit of efficient and sustainable desalination technologies has become not only a scientific imperative but also a societal necessity. This book, "Advances in Desalination Insights", is crafted to serve as a comprehensive guide and a source of inspiration for researchers, engineers, policymakers, and all those engaged in the noble quest of providing clean and accessible freshwater from the vast expanse of seawater.

The journey through the pages of this book commences with an exploration of solar-driven interfacial evaporation. Solar-driven interfacial evaporation represents a paradigm shift in desalination, leveraging the sun's power through advanced photothermal materials and innovative evaporator designs. As detailed in the opening chapter, remarkable strides have been made in enhancing the evaporation efficiency and salt resistance of SDIE systems, heralding a new era of green desalination possibilities. This sets the stage for subsequent discussions on the progress of heat-localized solar evaporation, which delves deeper into the mechanisms and optimizations of this technology, highlighting its potential to achieve high evaporation efficiencies and its application in solar stills.

The development path of solar thermal seawater desalination technology is comprehensively mapped. Traditional desalination methods have long been plagued by issues such as high energy consumption and environmental impacts. In contrast, solar-powered alternatives offer a glimmer of hope. This chapter dissects the various direct and indirect solar desalination techniques, analyzing their strengths and weaknesses and laying a roadmap for future advancements, focusing on enhancing efficiency, reducing costs, and promoting large-scale implementation.

The book focuses on membrane separation processes as we transition from solar-based technologies. These processes are the workhorses of modern desalination with various industry applications. The in-depth exploration of membrane principles, structures, materials, and preparation techniques provides a fundamental understanding of their role in desalination and paves the way for future innovations in this critical area.

Adsorbents for water desalination offer another promising avenue. With the global demand for freshwater surging, adsorption techniques have emerged as a viable option due to their potential for high salt removal efficiency and low environmental footprint. This chapter evaluates the latest adsorbent materials, from activated carbons to advanced nanomaterials like carbon nanotubes and graphene, shedding light on their properties and potential applications while also addressing the challenges associated with their synthesis and use.

Integrating machine learning in desalination is a rapidly evolving field, and our book dedicates a chapter to this exciting intersection. Machine learning is revolutionizing how we understand and optimize evaporation performance in porous media. Restructuring porous media structure information, establishing coefficient widespread adoption in economically challenged regions. Hence, the urgent pursuit of novel desalination technologies that are cost-effective, energy-efficient, eco-friendly, sustainable, and highly productive holds paramount significance in reconciling the desalination-energy consumption paradox and alleviating the global freshwater crisis. By vigorously investing in such innovations, we can pave the way for a more secure and sustainable water future.

Solar energy, an abundant and pristine resource, has been harnessed since antiquity, contributing immensely to the advancement of human civilization. Photothermal conversion, a pivotal mode of harnessing this energy, not only provides the power for the natural processes of hydrological cycles and atmospheric circulation but also offers a beacon of hope in addressing water scarcity. Solar-driven evaporation, leveraging solar energy exclusively as its power source, circumvents the need for fossil fuels, thereby reconciling the conflict between freshwater availability and energy consumption. According to the position of photothermal materials in liquid media, the early solar-driven evaporation systems can be divided into two categories: bottom-heating system and bulk heating system (**Figure 1a** and **b**) [6]. The bottom-heating system is a solar-driven evaporation system in which sunlight usually needs to pass through the bulk water to be absorbed by the photothermal materials at the bottom surface (**Figure 1a**). This evaporation system requires heating the entire water body, resulting in significant heat loss, and its evaporation efficiency is only 30–45% [7]. The bulk heating system is a solar-driven evaporation system that directly absorbs and converts solar energy through nanoparticles dispersed in the fluids (**Figure 1b**) and can generate steam without heating the bulk water to boiling point. However, this system also faces the challenges of difficult pumping, easy aggregation, and deposition of nanofluids [8, 9].

In recent years, researchers have devoted significant efforts toward the development of advanced and highly efficient solar-driven evaporation systems, notably the interfacial heating system (depicted in **Figure 1c**). A pivotal moment occurred in 2014 when Prof. Chen's team at MIT and Prof. Deng's team from Shanghai Jiao Tong University concurrently published groundbreaking articles in Nature Communications [10] and Small [11], introducing the novel concept of interfacial evaporation at the air-water interface, respectively. The solar-driven interfacial

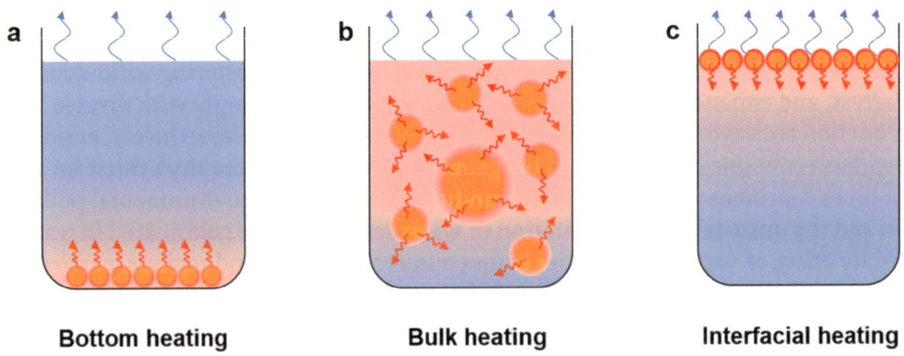

Figure 1.
Solar-driven evaporation through various forms of solar heating. (a) Bottom-heating system in which solar is absorbed by a solar absorber and converted into thermal energy to heat the bulk water from the bottom. (b) Bulk-heating system in which homogeneously dispersed solar absorbers (nanoparticles) convert incident solar photons into thermal energy to heat the bulk water. (c) Interfacial-heating system in which the solar-thermal conversion and evaporation are localized at the air-liquid interface.

evaporation (SDIE) system distinguishes itself from other solar-driven evaporation systems by precisely locating the conversion of solar energy into thermal energy at the air-liquid interface, only heating the water at the interface thereby reducing heat loss and achieving excellent evaporation performance and efficient photothermal conversion efficiency. Furthermore, the effective evaporation process inherent in SDIE systems keeps the surface temperature of the absorbing material relatively low, which in turn mitigates convective and radiative heat losses at the material's surface [6].

A conventional solar-driven interfacial evaporation system adopts a bilayer structure, as illustrated in **Figure 2**. The top layer comprises a solar-absorbing layer designed to efficiently capture and convert solar energy into heat, while simultaneously permitting the passage of steam through its surface without substantial contact with the underlying bulk water. The bottom layer serves a dual purpose: it restricts the conduction of heat generated by the photothermal materials into the bulk water, ensuring minimal heat loss, and provides mechanical stability and support for the solar-absorbing layer. Furthermore, this bottom layer necessitates a microporous structure featuring interconnected channels to efficiently supply water to the heated region, fostering rapid and stable evaporation rates.

In the past decade, this technology has experienced rapid development, with evaporation efficiency increasing to over 90% under conventional solar concentration, mainly enable by the continuous improvement and innovation of new photothermal materials and evaporator structural engineering [12–15]. The emergence of these new photothermal materials is mainly benefited from the continuous progress and development of modern nanotechnology in materials science. The vast majority of these photothermal materials has a light absorption rate of over 95% [16], which provides a guarantee for efficient solar-driven interface evaporation. Furthermore, a reasonable evaporator structure design has also made an important contribution to minimizing the system's heat loss. With the continuous development of solar-driven interfacial evaporation technology, it is possible to provide new insights for the application of the energy industry in modern society. Currently, solar-driven interface evaporation technology has achieved zero greenhouse gas emissions while providing clean water, making it an attractive advantage in addressing clean energy demand, water scarcity, and global warming issues.

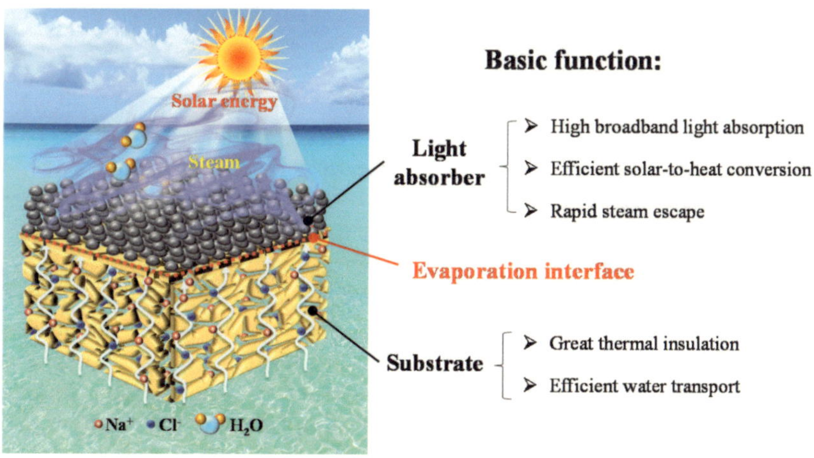

Figure 2.
The overall structural scheme of a typical bilayer interfacial solar-driven evaporation device.

This chapter provides a detailed introduction to the development of solar-driven interfacial evaporation technology, including the selection of various types of photothermal materials and supporting substrates and structural engineering strategies to improve thermal efficiency, and points out the opportunities and challenges faced by the further development of this technology.

2. Design principles to high evaporation performance

The evaporation performance of solar-driven interfacial evaporation systems is closely related to three factors, namely the solar absorption, thermal management and water transport [17]. It is worth noting that these three factors are closely related rather than independent of each other. The use of reasonable photothermal structure design to synergistically match these factors is intuitively important for optimizing the evaporation performance of the interfacial evaporation system. Therefore, this section first introduces the performance evaluation indicators of the interfacial evaporation system and then describes in detail three key strategies to improve evaporation performance: (i) efficient solar absorption; (ii) rapid and sufficient water transport; and (iii) excellent thermal management.

2.1 Evaluation indicators for evaporation performance

The performance of the solar-driven interfacial evaporation (SDIE) system is critically evaluated through two key indicators: evaporation rate and evaporation efficiency. The evaporation rate represents the mass variation of the SDIE system per unit time and unit area, which can be mathematically quantified using Eq. (1). In addition, evaporation efficiency measures the fraction of incoming solar flux that is effectively converted into stored thermal energy within the generated vapor, a calculation facilitated by Eq. (2). Both indicators provide valuable insights into the system's operational effectiveness and efficiency.

$$M = \frac{\Delta m}{S \cdot t} \tag{1}$$

$$\eta = \frac{M h_{eva}}{C_{opt} P_0} \tag{2}$$

where M refers to the evaporation rate of water, $\triangle m$ refers to the mass changes of the system, S is the projected area of the evaporator, t is the running time of the system, η refers to the evaporation efficiency, h_{eva} is the total liquid-vapor phase-change enthalpy (including the sensible heat), P_0 is the nominal solar irradiation value of 1 kW m^{-2}, and C_{opt} represents the optical concentration.

The theoretical limit of the evaporation rate of water under 1 kW m^{-2} illumination is 1.47 kg m^{-2} h^{-1} (3600 × 1 kW m^{-2}/2450 kJ kg^{-1} ≈ 1.47 kg m^{-2} h^{-1}). Nevertheless, numerous evaporator systems reported in literature exhibit evaporation rates surpassing this theoretical limit, leading to evaporation efficiencies that exceed 100% [18]. Consequently, when assessing the actual evaporation efficiency of a system, it is imperative to account for the environmental energy absorbed by the system and the variations in the enthalpy of water evaporation. Taking these factors into account, the precise evaporation efficiency of the system can be accurately calculated by

$$\eta' = \frac{(M - M_{\text{dark}})h_{act}}{C_{opt}P_0} \quad (3)$$

where M_{dark} refers to the evaporation rate of water in dark conditions and h_{act} is the verily liquid-vapor phase-change enthalpy.

2.2 Efficient solar absorption

In a solar-driven interfacial evaporation system, when sunlight illuminates on the photothermal materials, they are absorbed and converted into thermal energy, which is used to heat the water at the air-liquid interface, a process known as solar heat localization. Thus, photothermal materials with a wide spectrum of solar absorption and efficient conversion of solar energy into heat to drive water evaporation are the key factors in achieving efficient solar-driven interfacial evaporation [6]. So far, hundreds types of photothermal materials have been developed and explored, which can be classified into carbon-based materials, plasmonic particles, and semiconductor materials based on the differences in photothermal conversion mechanisms [19–22].

2.2.1 Carbon-based materials

Carbon based materials typically exhibit strong solar absorption abilities and can convert solar energy into thermal energy through lattice vibrations. As shown in **Figure 3a**, in carbon-based materials, loosely held electrons can be easily excited

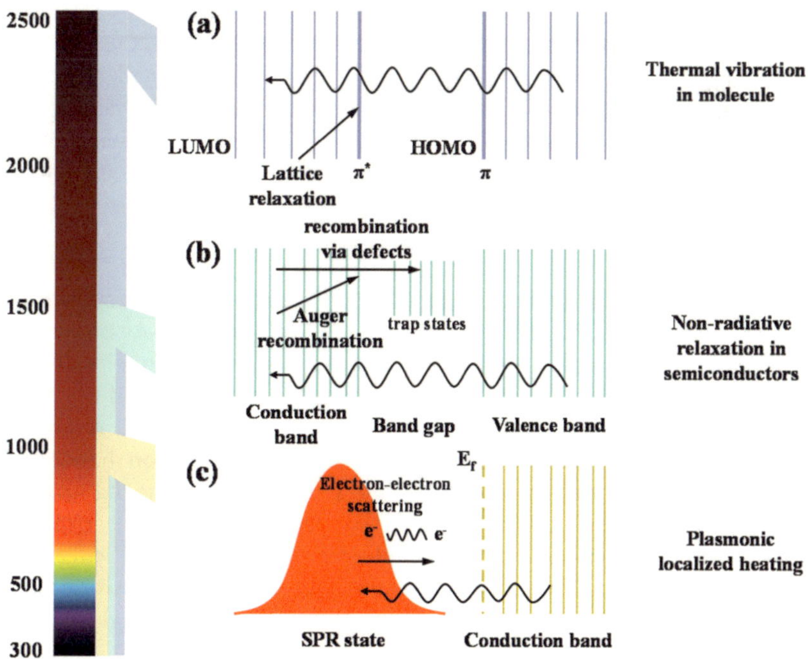

Figure 3.
The photothermal conversion mechanism and corresponding spectral absorption wavelength range of different types of photothermal materials. (a) Carbon-based materials. (b) Semiconductor materials. (c) Plasmonic particles.

from π orbitals to π* orbitals even with a small amount of energy input. When the photon energy of the incident light matches the possible electron transitions within the molecule, the excited electrons are lifted from the ground state (highest occupied molecular orbital, HOMO) to a higher energy orbital (lowest unoccupied molecular orbital), causing the excited electrons to relax back to the ground state and release heat [23, 24].

Because of their high solar absorption across a wide range of wavelengths (range from 250 nm to 2500 nm) and relatively low cost, carbon-based materials have been widely used in the field of solar-driven interfacial evaporation, such as graphene [25, 26], graphene oxide (GO) [27], reduced graphene oxide (rGO) [28], carbon nanotubes [29–31], and carbon black [32]. In addition, due to the strong stability and processability of carbon-based materials, they can be well integrated with porous substrates to enhance their solar absorption abilities and form stable evaporation structures.

2.2.2 Semiconductor materials

Recently, low-cost and low-toxicity semiconductors have emerged as a new type of photothermal material for solar-driven interfacial evaporation. Compared to carbon-based materials, there are fewer types of semiconductor materials, mainly metal oxide compounds, such as hydrogenated black titania [33], Fe_3O_4 [34], oxygen-deficient MoO_3 quantum dots [35], and bimetal oxides [36, 37]. As shown in **Figure 3b**, semiconductor materials have a high absorption for light in the wavelength range of 300–1500 nm [38]. In semiconductor materials, when the material absorbs light with energy similar to the bandgap, electron hole pairs are generated. In narrow bandgap semiconductors, due to the fact that the energy of most photons in the incident sunlight is higher than the energy of the bandgap, electron hole pairs above the bandgap will be generated. These electron hole pairs above the bandgap will then relax to the band edge, converting additional energy into heat through a thermalization process. In sharp contrast, when electron hole pairs recombine near the bandgap edge, most of the absorbed light energy is re-emitted in the form of photons in the wide bandgap semiconductor, resulting in much lower photothermal conversion efficiency [39].

2.2.3 Plasmonic particles

Plasmonic particle absorbers refer to various metal nanoparticles, such as gold (Au) [40, 41], silver (Ag) [41, 42], germanium (Ge) [43], and Au-Ag bimetals [41], etc., which have high solar absorption for light with wavelengths ranging from 300 nm to 1100 nm (**Figure 3c**) [44]. The photothermal conversion of the plasmonic particles is caused by the localized surface plasmon resonance (LSPR) effect of metals [45]. LSPR is a resonant photon-induced charge coherent oscillation that occurs when the photon frequency matches the natural frequency of the metal surface electrons. When sunlight shines on the surface of plasma, it excites electrons from occupied to unoccupied states, forming hot electrons (phonons) and redistributing energy through electron and electron-phonon scattering, causing the local surface temperature of the metal to rapidly rise. Then, the lattice cools through coupling and dissipates heat to the surrounding fluid.

In the early stage of research, to explore the solar evaporation performance using plasma nanoparticles, a self-assembly Au nanoparticles thin film with wide solar absorption spectrum was fabricated and placed at the air-liquid interface [46]. There is a significant heat loss from Au NP film to the bulk water because of their direct contact. The photothermal conversion efficiency of the film system is only about 44%, but this is still more than twice as high as the efficiency of the gold NPs suspension system. To further improve the efficiency of photothermal conversion, plasmonic particle absorbers were deposited on the porous substrates (such as rough paper [19], wood [47–50], and multi-channeled anodic aluminum oxide (AAO) structure [51]) to reduce the heat loss of system. Due to the complex micro-interface morphology of porous substrates, multiple reflections and refractions of sunlight have been achieved, resulting in an ultra-high solar absorption of approximately 99% for plasmonic particle absorbers [41, 42].

2.3 Continuous and efficient water transport

Maintaining sufficient and continuous water supply through excellent water pathway design is the fundamental guarantee for achieving efficient and sustainable solar evaporation. In conventional solar evaporators, porous substrates typically have 3D random and interconnected porous structures that can continuously transport water from the underlaying water to the evaporation interface (**Figure 4a**). For example, it is by using natural porous materials [55–57] or commercial porous sponges [58] with rich internal porous structures as substrates to ensure sufficient water supply, and the interfacial evaporation system can be prepared by surface carbonization or loading photothermal materials. However, the thermal conductivity of these substrate materials increases after wetting, which leads to an increase in thermal loss from the evaporation interface to the bulk water, thereby weakening the thermal localization ability of the system.

To alleviate the problem of excessive thermal conductivity loss, a two-dimensional water pathway structure (**Figure 4b**) is formed by wrapping the water transport material on the surface of an insulation material with low thermal conductivity [47]. The evaporation system with this two-dimensional water pathway physically isolates the absorber from the bulk water through insulation materials, greatly suppressing parasitic heat loss. Compared with the overall water supply in conventional designs, the limited two-dimensional water channel structure reduces the dimensionality and volume of the water pathway, thereby minimizing the heat dissipation of the evaporation system to the bulk water [55]. Furthermore, as shown in **Figure 4c**, the dimension of the water pathway structure is reduced from 2D to 1D, forming a transport mechanism similar to water being transported from the mushroom stem to the cap [55]. And the water can be directly transported to the photothermal material through the

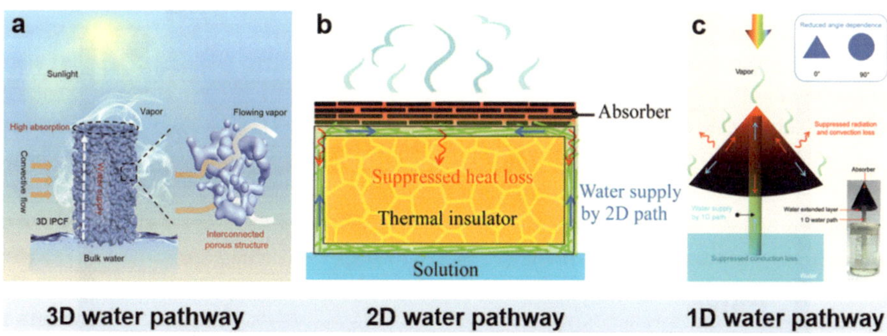

Figure 4.
The three types of evaporators with different water transport paths [52–54].

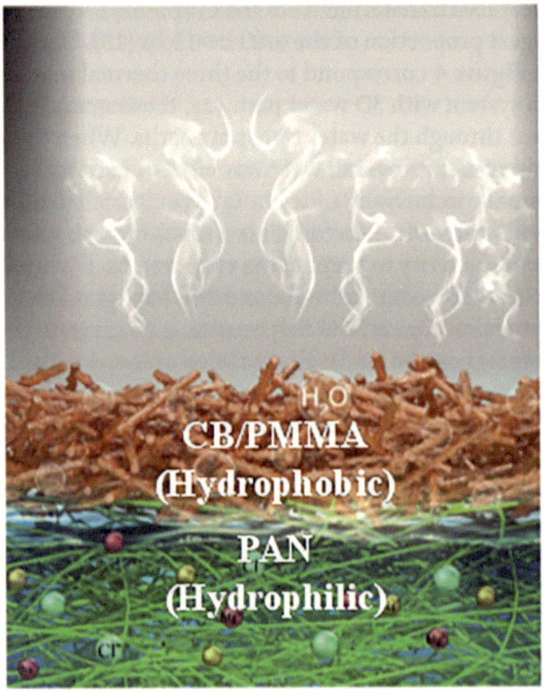

Figure 6.
Janus structure for enhancing salt resistance [63].

the hydrophobic and the hydrophilic layer, and they can timely diffuse back to the brine through the hydrophilic bottom layer, with a higher evaporation rate than the all-hydrophobic evaporator [63, 64].

During water transport, salt ions are always carried and raised to the evaporation region, which is the prerequisite for salt precipitation. Thus, separating water and salt ions during water transport is an imaginative method for enhancing salt resistance. Donnan effect can be utilized to repel the anion/cation from the charged hydrogel evaporator during the transport (**Figure** 7). Because the water supply layer of some hydrogel evaporators has a fixed charge, the co-ions are almost kept away from the

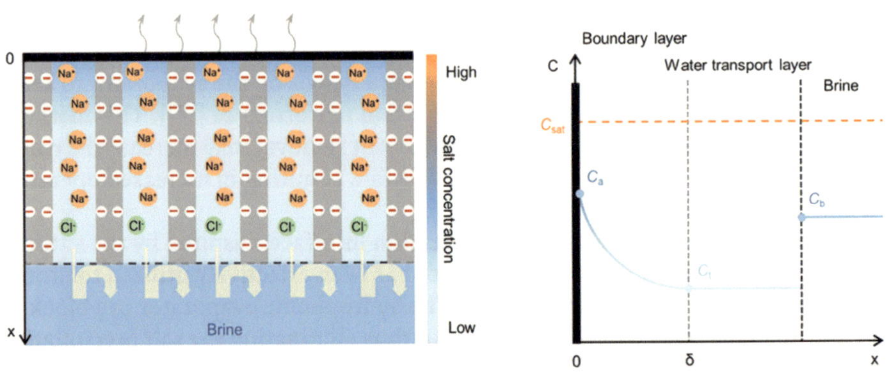

Figure 7.
Donnan effect for separating ions and water [65].

evaporator, and the counter-ions that enter the hydrogel with water are absorbed by the charged hydrogel due to electric neutrality, consequently separating the anion and cation and avoiding salt precipitation [65].

Actually, it is well recognized that the simplest and the most feasible method for salt resistance is to enhance the diffusion and convection of ions by accelerating the water transport. Because of the differences in the salt concentration and density between the evaporator and the bulk water, salt ions will flow back to the bulk water due to diffusion and convection. For example, some evaporators can dissolve the salt crystals when they do not work at night [66]. Therefore, speeding up diffusion and convection is an effective way to block salts (**Figure 8**). Normally, improving water transport can promote this process. For example, enhancing the hydrophilicity and designing additional water transport channels can improve the water supply, and the evaporators can stably work even under focus radiation and high salinity brine [67, 68]. Besides, according to the Fick's law, a short mass-transfer distance also causes a quick salt ion migration. Thus, reducing the tortuosity of the porous substance can shorten the distance of the ion migration and enhance the diffusion, consequently improving the salt resistance [69]. However, this method always promotes heat loss and reduces the evaporation rate because of the coupling between heat and mass transfer.

Besides the capillary force, incorporating other forces can further improve the water supply. For example, the Marangoni flow driven by differences in surface tension can quickly transport salt ions to the bulk water (**Figure 9a**) [70]. Moreover, the evaporator with a liquid level difference can form a directional flow to bring the ions away from the evaporation regions (**Figure 9b**) [71]. Besides, external electrical force and imbalanced torque are also utilized to migrate the ions and enhance the salt resistance [72, 73].

Except for the above strategies, contactless evaporation is a general and elaborate method for permanently blocking salt precipitation (**Figure 10**) [74]. The pivotal component of this method is an absorber/emitter unit above the water, which can convert the solar flux to mid-infrared radiation. Meanwhile, a thin water layer serving as the evaporator can efficiently absorb the mid-infrared radiation and evaporate (most within 100 μm layer). Thus, the solar absorber is completely and physically separated from the evaporator, and thus no salt will precipitate on this solar absorber. But this method always has a lower evaporation rate.

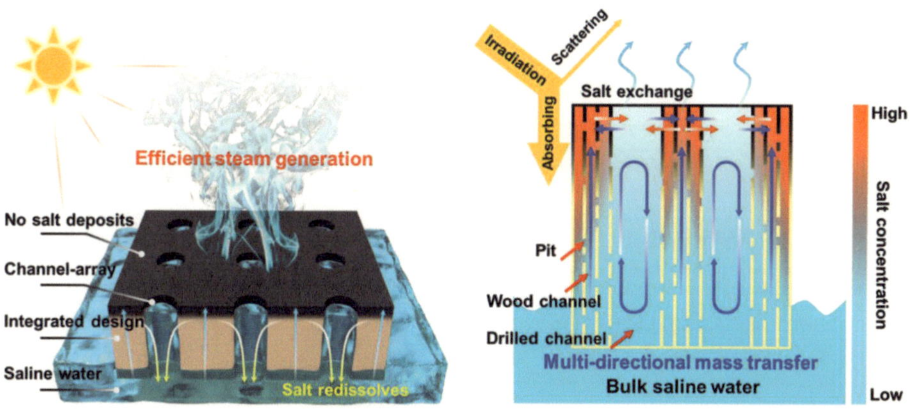

Figure 8.
Improved water supply for enhancing salt exchange and salt resistance [67].

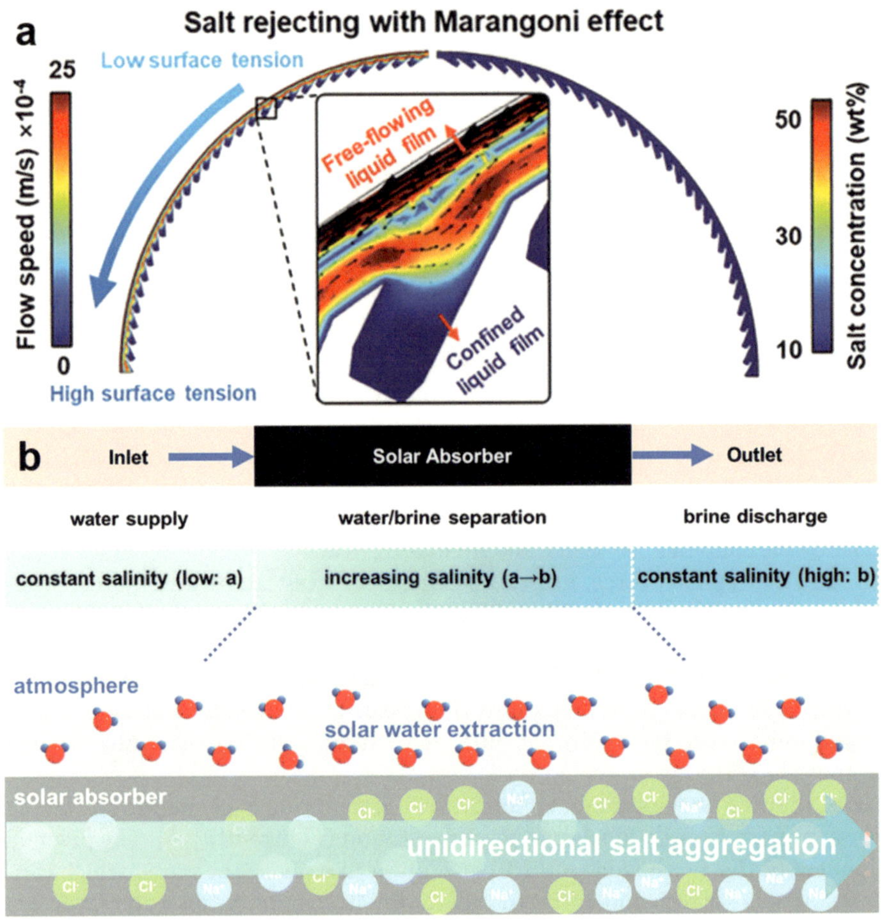

Figure 9.
Incorporating Marangoni effect [70] (a) and one-way flow [71] (b) for rapid salt ion migration.

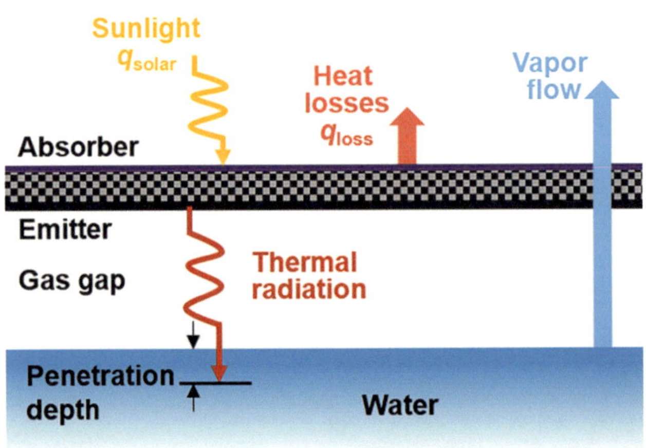

Figure 10.
The mechanism of the contactless evaporator [75].

Figure 11.
One-way water transport channel for ZLD [76].

The above strategies can significantly enhance the salt resistance, but the salts always flow back to the brine, which causes the increasing salinity of the brine and is not beneficial for mineral recycle. Zero liquid discharge (ZLD) has been recognized as an efficient strategy to simultaneously harvest freshwater and salts even in high-salinity brine [76]. By purposively depositing the salt crystals on the edge of the evaporator and separating the photothermal surface from the salt deposit region, the evaporator can work steadily and recycle salts. For example, the evaporator with an ingenious one-way water transport channel from the central to the edge of the evaporator achieves a radially reduced water supply (**Figure 11**). Thus, the salt concentration increases from the central to the edge, and the salt crystals finally form at the edge of the evaporator, which does not affect the evaporation. Moreover, the salt crystals precipitating on the pre-wetted evaporator prefer to automatically be dropped by gravity force, consequently finishing the salt recycle. Thus, this method shows potential for water and mineral harvest and extends the application of SDIE.

In general, many efficient and ingenious salt-resistant methods have been proposed, and the salt resistance of the evaporator has been remarkably enhanced. Based on these methods, the evaporators can stably and continuously evaporate even under high-salinity brine. Thus, this technology shows potential for industrial desalination and stable access to fresh water.

4. Conclusions and outlooks

Solar-driven interfacial evaporation shows green, sustainable, and efficient character, which is a promising solution for alleviating the freshwater crisis. Owing to the researchers' unremitting endeavor, the past decade has witnessed the rapid and remarkable development in the field of the solar-driven interfacial evaporation, which significantly enhances the evaporation rate and the salt resistance of the evaporator. Besides, it also can be utilized in the fields of electrical generation, wastewater purification, and fuel generation, indicating its extensive application. However, there are also some key issues limiting the scalable and commercial application, summarized as follows:

1. Deeper comprehension in the field of evaporation process. Though the evaporation rate has been remarkably enhanced, the understanding of this multi-field coupling evaporation process is limited. For example, the reason for the reduced evaporation enthalpy is not clear, and the energy transfer and conversion in the multifunction system are not revealed. Deep comprehension of SDIE is conducive to designing the evaporator with better performance. With the help of kinds of numerical simulations, the mechanism behind the evaporation may be uncovered.

2. Long-term stability. Due to the complexity of the actual seawater, it is necessary to consider the potential failure of the evaporator caused by kinds of organics, microorganisms, and salts. Thus, reasonable design for the evaporator with high resistance to contamination can efficiently prolong the longevity of the system and lower the cost.

3. Limited condensate collection. The evaporation rate in the open system has been remarkably enhanced, even breaking the theory limit. After integrating a transparent cover, this closed system can obtain freshwater. Different from the open system, the vapor will condense on the cover, which causes high optical loss. Then, limited condensation makes the humidity in the closed system quite high, consequently decreasing the evaporation. Eventually, the collection rate in the closed system is quite low, which is not beneficial for obtaining freshwater. Therefore, improving the collection rate by reasonable design is more meaningful for application.

In brief, it is necessary to deeply investigate the solar-driven interfacial evaporation, and this innovative and sustainable technology can play a more important role in efficiently obtaining freshwater.

Author details

Yunqi Li†, Haixiang Feng†, Yu Qiu and Qing Li*
School of Energy Science and Engineering, Central South University, Changsha, Hunan, China

*Address all correspondence to: qingli@csu.edu.cn

† These authors contributed equally to this work.

References

[1] Gleick PH, Cooley H. Freshwater scarcity. Annual Review of Environment and Resources. 2021;**46**(1):319-348

[2] Salehi M. Global water shortage and potable water safety; Today's concern and tomorrow's crisis. Environment International. 2022;**158**:106936

[3] Kurihara M. Seawater reverse osmosis desalination. Membranes. 2021;**11**(4):243

[4] Ali E, Orfi J, AlAnsary H, Baakeem S, Alsaadi AS, Ghaffour N. Advanced structures of reversal multi-stage flash desalination. Desalination. 2024;**571**:117095

[5] Xu H, Ji X, Wang L, Huang J, Han J, Wang Y. Performance study on a small-scale photovoltaic electrodialysis system for desalination. Renewable Energy. 2020;**154**:1008-1013

[6] Tao P, Ni G, Song C, Shang W, Wu J, Zhu J, et al. Solar-driven interfacial evaporation. Nature Energy. 2018;**3**(12):1031-1041

[7] Kabeel AE, El-Agouz SA. Review of researches and developments on solar stills. Desalination. 2011;**276**(1-3):1-12

[8] Jin H, Lin G, Bai L, Zeiny A, Wen D. Steam generation in a nanoparticle-based solar receiver. Nano Energy. 2016;**28**:397-406

[9] Otanicar TP, Phelan PE, Prasher RS, Rosengarten G, Taylor RA. Nanofluid-based direct absorption solar collector. Journal of Renewable and Sustainable Energy. 2010;**2**(3):033102

[10] Ghasemi H, Ni G, Marconnet AM, Loomis J, Yerci S, Miljkovic N, et al. Solar steam generation by heat localization. Nature Communications. 2014;**5**:4449

[11] Wang Z, Liu Y, Tao P, Shen Q, Yi N, Zhang F, et al. Bio-inspired evaporation through plasmonic film of nanoparticles at the air-water interface. Small. 2014;**10**(16):3234-3239

[12] Wu X, Chen GY, Owens G, Chu D, Xu H. Photothermal materials: A key platform enabling highly efficient water evaporation driven by solar energy. Materials Today Energy. 2019;**12**:277-296

[13] Zhu L, Gao M, Peh CKN, Ho GW. Recent progress in solar-driven interfacial water evaporation: Advanced designs and applications. Nano Energy. 2019;**57**:507-518

[14] Dao V-D, Vu NH, Yun S. Recent advances and challenges for solar-driven water evaporation system toward applications. Nano Energy. 2020;**68**:104324

[15] Zhu L, Gao M, Peh CKN, Ho GW. Solar-driven photothermal nanostructured materials designs and prerequisites for evaporation and catalysis applications. Materials Horizons. 2018;**5**(3):323-343

[16] Kashyap V, Ghasemi H. Solar heat localization: Concept and emerging applications. Journal of Materials Chemistry A. 2020;**8**(15):7035-7065

[17] Chen C, Kuang Y, Hu L. Challenges and opportunities for solar evaporation. Joule. 2019;**3**(3):683-718

[18] Liu H, Huang Z, Liu K, Hu X, Zhou J. Interfacial solar-to-heat conversion for desalination. Advanced Energy Materials. 2019;**9**(21):1900310

[19] Gan Q, Zhang T, Chen R, Wang X, Ye M. Simple, low-dose, durable, and carbon-nanotube-based floating solar still for efficient desalination and purification. ACS Sustainable Chemistry & Engineering. 2019;7(4):3925-3932

[20] Gao M, Zhu L, Peh CK, Ho GW. Solar absorber material and system designs for photothermal water vaporization towards clean water and energy production. Energy & Environmental Science. 2019;**12**(3):841-864

[21] Li X, Yao Z, Wang J, Li D, Yu K, Jiang Z. A novel flake-like Cu7S4 solar absorber for high-performance large-scale water evaporation. ACS Applied Energy Materials. 2019;**2**(7):5154-5161

[22] Lin K, Chen R, Zhang L, Shen W, Zang D. Enhancing water evaporation by interfacial silica nanoparticles. Advanced Materials Interfaces. 2019;**6**(16):1900369

[23] Qiblawey HM, Banat F. Solar thermal desalination technologies. Desalination. 2008;**220**(1-3):633-644

[24] Vélez-Cordero JR, Hernández-Cordero J. Heat generation and conduction in PDMS-carbon nanoparticle membranes irradiated with optical fibers. International Journal of Thermal Sciences. 2015;**96**:12-22

[25] Yang J, Pang Y, Huang W, Shaw SK, Schiffbauer J, Pillers MA, et al. Functionalized graphene enables highly efficient solar thermal steam generation. ACS Nano. 2017;**11**(6):5510-5518

[26] Kim K, Yu S, An C, Kim S-W, Jang J-H. Mesoporous three-dimensional graphene networks for highly efficient solar desalination under 1 sun illumination. ACS Applied Materials & Interfaces. 2018;**10**(18):15602-15608

[27] Liu M, Zhang S, Yuan T, Bao C. Low-cost, scalable, and durable coal-based composite aerogel beads solar evaporator for efficient seawater desalination and wastewater purification. Desalination. 2023;**550**:116401

[28] Meng F, Zhang Y, Zhang S, Ju B, Tang B. Rational design of biomass-derived composite aerogels for solar-driven seawater desalination and sewage treatment. Industrial & Engineering Chemistry Research. 2022;**61**(27):9763-9773

[29] Ma X, Tian N, Wang G, Wang W, Miao J, Fan T. Biomimetic vertically aligned aerogel with synergistic photothermal effect enables efficient solar-driven desalination. Desalination. 2023;**550**:116397

[30] Wang Y, Li W, Wei Y, Chen Q. Recyclable monolithic Vitrimer foam for high-efficiency solar-driven interfacial evaporation. ACS Applied Materials & Interfaces. 2023;**15**(11):14379-14387

[31] Liu K, Zhang W, Cheng H, Luo L, Wang B, Mao Z, et al. A nature-inspired monolithic integrated cellulose aerogel-based evaporator for efficient solar desalination. ACS Applied Materials & Interfaces. 2021;**13**(8):10612-10622

[32] Zhang W, Zhang G, Ji Q, Liu H, Liu R, Qu J. Capillary-flow-optimized heat localization induced by an air-enclosed three-dimensional hierarchical network for elevated solar evaporation. ACS Applied Materials & Interfaces. 2019;**11**(10):9974-9983

[33] Li Z, Zhang J, Zang S, Yang C, Liu Y, Jing F, et al. Engineering controllable water transport of biosafety cuttlefish juice solar absorber toward remarkably enhanced solar-driven gas-liquid interfacial evaporation. Nano Energy. 2020;**73**:104834

[34] Song L, Zhang X-F, Wang Z, Zheng T, Yao J. Fe_3O_4/polyvinyl alcohol decorated delignified wood evaporator for continuous solar steam generation. Desalination. 2021;**507**:115024

[35] Ding D, Huang W, Song C, Yan M, Guo C, Liu S. Non-stoichiometric MoO_{3-x} quantum dots as a light-harvesting material for interfacial water evaporation. Chemical Communications. 2017;**53**(50):6744-6747

[36] Chen R, Wu Z, Zhang T, Yu T, Ye M. Magnetically recyclable self-assembled thin films for highly efficient water evaporation by interfacial solar heating. RSC Advances. 2017;7(32):19849-19855

[37] Shi Y, Li R, Jin Y, Zhuo S, Shi L, Chang J, et al. A 3D Photothermal structure toward improved energy efficiency in solar steam generation. Joule. 2018;**2**(6):1171-1186

[38] Ye M, Jia J, Wu Z, Qian C, Chen R, O'Brien PG, et al. Synthesis of black TiOx nanoparticles by Mg reduction of TiO_2 nanocrystals and their application for solar water evaporation. Advanced Energy Materials. 2016;**7**(4):1601811

[39] Chen X, Liu L, Yu PY, Mao SS. Increasing solar absorption for Photocatalysis with black hydrogenated titanium dioxide nanocrystals. Science. 2011;**331**(6018):746-750

[40] Wang Y, Ma H, Yu J, Li J, Xu N, Zhu J, et al. All-dielectric insulated 3D Plasmonic nanoparticles for enhanced self-floating solar evaporation under one Sun. Advanced Optical Materials. 2023;**11**(7):2201907

[41] Chen M, He Y, Huang J, Zhu J. Synthesis and solar photo-thermal conversion of Au, Ag, and Au-Ag blended plasmonic nanoparticles. Energy Conversion and Management. 2016;**127**:293-300

[42] Chen S, Sun Z, Xiang W, Shen C, Wang Z, Jia X, et al. Plasmonic wooden flower for highly efficient solar vapor generation. Nano Energy. 2020;**76**:104998

[43] Sun W, Zhong G, Kübel C, Ali FM, Qian C, Wang L, et al. Size-tunable photothermal germanium nanocrystals. Angewandte Chemie International Edition. 2017;**56**(22):6329-6334

[44] de Aberasturi DJ, Serrano-Montes AB, Liz-Marzán LM. Modern applications of plasmonic nanoparticles: From energy to health. Advanced Optical Materials. 2015;**3**(5):602-617

[45] Webb JA, Bardhan R. Emerging advances in nanomedicine with engineered gold nanostructures. Nanoscale. 2014;**6**(5):2502

[46] Zhou L, Tan Y, Ji D, et al. Self-assembly of highly efficient, broadband plasmonic absorbers for solar steam generation. Science Advances. 2016;**2**(4):e1501227

[47] Li T, Liu H, Zhao X, Chen G, Dai J, Pastel G, et al. Scalable and highly efficient mesoporous wood-based solar steam generation device: Localized heat, rapid water transport. Advanced Functional Materials. 2018;**28**(16):1707134

[48] Zhang X-F, Wang Z, Song L, Feng Y, Yao J. Chinese ink enabled wood evaporator for continuous water desalination. Desalination. 2020;**496**:114727

[49] Li L, He N, Yang S, Zhang Q, Zhang H, Wang B, et al. Strong tough hydrogel solar evaporator with wood skeleton construction enabling ultra-durable brine desalination. EcoMat. 2022;**5**(1):e12280

[50] Li Y, Li Q, Qiu Y, Feng H. High-efficiency wood-based evaporators for solar-driven interfacial evaporation. Solar Energy. 2022;**244**:322-330

[51] Traver E, Karaballi RA, Monfared YE, Daurie H, Gagnon GA, Dasog M. TiN, ZrN, and HfN nanoparticles on nanoporous aluminum oxide membranes for solar-driven water evaporation and desalination. ACS Applied Nano Materials. 2020;**3**(3):2787-2794

[52] Li J, Wang X, Lin Z, Xu N, Li X, Liang J, et al. Over 10 kg m^{-2} h^{-1} evaporation rate enabled by a 3D interconnected porous carbon foam. Joule. 2020;**4**(4):928-937

[53] Li X, Xu W, Tang M, Zhou L, Zhu B, Zhu S, et al. Graphene oxide-based efficient and scalable solar desalination under one sun with a confined 2D water path. Proceedings of the National Academy of Sciences. 2016;**113**(49):13953-13958

[54] Li X, Lin R, Ni G, Xu N, Hu X, Zhu B, et al. Three-dimensional artificial transpiration for efficient solar waste-water treatment. National Science Review. 2018;**5**(1):70-77

[55] Xu N, Hu X, Xu W, Li X, Zhou L, Zhu S, et al. Mushrooms as efficient solar steam-generation devices. Advanced Materials. 2017;**29**(28):1606762

[56] Xue G, Liu K, Chen Q, Yang P, Li J, Ding T, et al. Robust and low-cost flame-treated wood for high-performance solar steam generation. ACS Applied Materials & Interfaces. 2017;**9**(17):15052-15057

[57] Zhang H, Li L, Jiang B, Zhang Q, Ma J, Tang D, et al. Highly thermally insulated and superhydrophilic corn straw for efficient solar vapor generation. ACS Applied Materials & Interfaces. 2020;**12**(14):16503-16511

[58] Liu Y, Liu Z, Huang Q, Liang X, Zhou X, Fu H, et al. A high-absorption and self-driven salt-resistant black gold nanoparticle-deposited sponge for highly efficient, salt-free, and long-term durable solar desalination. Journal of Materials Chemistry A. 2019;**7**(6):2581-2588

[59] Xu K, Wang C, Li Z, Wu S, Wang J. Salt mitigation strategies of solar-driven interfacial desalination. Advanced Functional Materials. 2020;**31**(8):2007855

[60] Ren H, Tang M, Guan B, Wang K, Yang J, Wang F, et al. Hierarchical graphene foam for efficient omnidirectional solar–thermal energy conversion. Advanced Materials. 2017;**29**(38):1702590

[61] Sheng M, Yang Y, Bin X, Zhao S, Pan C, Nawaz F, et al. Recent advanced self-propelling salt-blocking technologies for passive solar-driven interfacial evaporation desalination systems. Nano Energy. 2021;**89**:106468

[62] Zhao J, Yang Y, Yang C, Tian Y, Han Y, Liu J, et al. A hydrophobic surface enabled salt-blocking 2D Ti3C2MXene membrane for efficient and stable solar desalination. Journal of Materials Chemistry A. 2018;**6**(33):16196-16204

[63] Xu W, Hu X, Zhuang S, Wang Y, Li X, Zhou L, et al. Flexible and salt resistant Janus absorbers by electrospinning for stable and efficient solar desalination. Advanced Energy Materials. 2018;**8**(14):1702884

[64] Yang Y, Zhao H, Yin Z, Zhao J, Yin X, Li N, et al. A general salt-resistant hydrophilic/hydrophobic nanoporous double layer design for efficient

and stable solar water evaporation distillation. Materials Horizons. 2018;**5**(6):1143-1150

[65] Zhao W, Gong H, Song Y, Li B, Xu N, Min X, et al. Hierarchically designed salt-resistant solar evaporator based on Donnan effect for stable and high-performance brine treatment. Advanced Functional Materials. 2021;**31**(23):2100025

[66] Zhang Q, Li L, Jiang B, et al. Flexible and mildew-resistant wood-derived aerogel for stable and efficient solar desalination. ACS Applied Materials & Interfaces. 2020;**12**(25):28179-28187

[67] Kuang Y, Chen C, He S, Hitz EM, Wang Y, Gan W, et al. A high-performance self-regenerating solar evaporator for continuous water desalination. Advanced Materials. 2019;**31**(23):1900498

[68] Fang Q, Li T, Lin H, Jiang R, Liu F. Highly efficient solar steam generation from activated carbon fiber cloth with matching water supply and durable fouling resistance. ACS Applied Energy Materials. 2019;**2**(6):4354-4361

[69] Li Y, Li Q, Qiu Y, Feng H, Deng R. A novel aerogel-based solar evaporator with triple-layered low-tortuosity pore structures for ultra-high salt resistance. Solar RRL. 2024;**8**(18):2400418

[70] Zou M, Zhang Y, Cai Z, Li C, Sun Z, Yu C, et al. 3D printing a biomimetic bridge-arch solar evaporator for eliminating salt accumulation with desalination and agricultural applications. Advanced Materials. 2021;**33**(34):2102443

[71] Zhang Y, Zhang H, Xiong T, Qu H, Koh JJ, Nandakumar DK, et al. Manipulating unidirectional fluid transportation to drive sustainable solar water extraction and brine-drenching induced energy generation. Energy & Environmental Science. 2020;**13**(12):4891-4902

[72] Fu S, Chen Y, Tao Q, Li C, Liu X, He D. Solar-driven interfacial desalination actively adjusted by the electric field. Separation and Purification Technology. 2023;**327**:124987

[73] Xia Y, Li Y, Yuan S, Kang Y, Jian M, Hou Q, et al. A self-rotating solar evaporator for continuous and efficient desalination of hypersaline brine. Journal of Materials Chemistry A. 2020;**8**(32):16212-16217

[74] Bian Y, Ye Z, Zhao G, Tang K, Teng Y, Chen S, et al. Enhanced contactless salt-collecting solar desalination. ACS Applied Materials & Interfaces. 2022;**14**(29):34151-34158

[75] Cooper TA, Zandavi SH, Ni GW, Tsurimaki Y, Huang Y, Boriskina SV, Chen G. Contactless steam generation and superheating under one sun illumination. Nature Communications. 2018;**9**(1):5086

[76] Xia Y, Hou Q, Jubaer H, Li Y, Kang Y, Yuan S, et al. Spatially isolating salt crystallisation from water evaporation for continuous solar steam generation and salt harvesting. Energy & Environmental Science. 2019;**12**(6):1840-1847

Chapter 2

Progress in Heat-Localized Solar Evaporation for Sustainable Desalination

Xiao Luo and Xiaokun Gu

Abstract

In recent years, the localized solar thermal evaporation technology for solar seawater desalination systems has gained widespread attention, primarily due to its ability to achieve over 80% evaporation efficiency under a single sun intensity. As research progresses, solar seawater desalination systems based on this technology have continually made breakthroughs in freshwater production. This chapter introduces the application of interfacial evaporation technology in solar seawater desalination systems, including the development and improvement of interfacial evaporation solar stills. First, the theoretical analysis methods and mathematical models of interfacial evaporation solar stills are discussed. Secondly, based on the interfacial evaporation structure, the performance optimization methods of solar stills are explored. Finally, the application of multistage waste heat recovery technology in interfacial evaporation solar stills is discussed. This chapter provides a broader scope for the application of localized solar evaporation in seawater desalination systems.

Keywords: heat-localized solar evaporation, interfacial evaporation, solar desalination, solar energy, heat and mass transfer

1. Introduction

With the rapid growth of the global population and social economy, water scarcity has become one of the major crises facing human society. Solar-driven seawater evaporation desalination is considered a low-carbon and environmentally friendly technology that can alleviate water scarcity issues [1, 2]. In recent years, with the development of interfacial evaporation structures that can achieve high evaporation rates [3–5], photothermal seawater desalination has once again become a research hotspot in the field of seawater desalination.

Unlike traditional solar stills that use volumetric heating and bottom heating methods, interfacial evaporation structures are composed of light-absorbing materials for localized solar energy and adiabatic water supply materials to reduce heat loss and ensure adequate water supply [4–8]. By reducing the heat capacity of the seawater to be evaporated, these structures can improve evaporation rates under normal light intensity (approximately 1 $kW{\cdot}m^{-2}$) without the need for light concentration [9–12].

Currently, most interfacial evaporation structures can exceed evaporation rates of 1.2 $kg{\cdot}m^{-2}{\cdot}h^{-1}$ under light intensity of 1 $kW{\cdot}m^{-2}$ [10, 12–16].

In scenarios with limited building area and no additional energy supply, solar stills are among the most widely used solar desalination devices due to their simple structure, ease of production, operation, and maintenance [17–21]. However, the freshwater production efficiency of traditional solar stills is relatively low. This efficiency, defined as the ratio of the energy released by the condensation of water vapor into freshwater to the amount of solar radiation input, has an ideal efficiency of only 60%, with actual efficiency difficult to exceed 50% [17, 22–25]. With the development of interfacial evaporation structures, solar stills based on these structures are expected to achieve breakthroughs in freshwater production efficiency and daily freshwater yield [26, 27]. However, there is still significant room for improvement in the research on interfacial evaporation structures in seawater desalination systems.

Regarding the optimization methods for solar stills, previous studies have proposed relevant strategies. One possible strategy is to further improve the evaporation efficiency of the interfacial evaporation structure, which is believed to be beneficial for increasing water yield. Studies have found that the water production efficiency of single-stage solar stills with interfacial evaporation structures is 25% higher than that of single-stage solar stills using traditional heating methods [28]. However, experimental results show that the interfacial evaporation structure is limited by environmental factors in closed spaces and cannot effectively improve the water production efficiency of solar stills. Another possible strategy is to optimize the design of interfacial evaporative solar stills by referring to methods used in the past to optimize conventional solar stills. Many researchers have enhanced condensation by strengthening convective heat transfer outside the glass cover, which has shown good effects on solar stills with volumetric heating, increasing efficiency by about 10% compared to solar stills without external enhanced heat transfer [29]. In recent years, significant research has been conducted on the theoretical analysis and performance optimization of interfacial evaporation stills. Through the exploration of methods to enhance the parameters of these stills, improvements in performance have been achieved [30].

In addition, waste heat utilization is another effective measure to optimize the performance of solar stills. Traditional solar stills improve their performance by utilizing waste heat through multieffect distillation but this often increases the complexity of the device and affects the light intensity absorbed by the evaporation layer [31, 32]. Inspired by traditional multieffect condensation solar stills, researchers have designed a thin-layer evaporation multistage condensation system suitable for interface evaporation structure [33]. A hydrophobic membrane such as polytetrafluoroethylene is used as a separation membrane between the seawater layer and the freshwater layer, allowing water vapor to enter the freshwater layer through the membrane for condensation and collection and then transferring the latent heat of condensation to the next seawater layer [34]. This not only avoids seawater pollution to freshwater but also achieves multistage condensation effect. Therefore, in the research of interface evaporation distillers, it is necessary to consider the significant increase in freshwater production by multistage condensation systems and propose more efficient and rational distiller designs that utilize waste heat.

This chapter reviews the research progress of interface evaporation solar distillers, starting from the theoretical calculation model of distillers; analyzes the heat and mass transfer process of the interface evaporation structure of solar distillers; and studies the optimization direction of interface evaporation solar distillers. Through literature analysis, key parameters affecting the water production efficiency of the interface evaporative solar distiller are determined, such as the ratio of the illumination area to the evaporation area of the internal structure of the solar distiller, the thermal insulation performance of the solar still, the emissivity of the absorbing material, external and internal convective heat transfer terms, and the installation of waste heat utilization devices. The optimization direction of the interfacial evaporation solar still is analyzed.

2. Theoretical models of the interfacial evaporation solar stills

To explore the mechanisms of solar seawater desalination systems based on interfacial evaporation structures, researchers have established a detailed heat and mass transfer model for solar stills according to the law of energy conservation. The heat transfer process is illustrated in **Figure 1a**. Under transient conditions, two energy balance equations can be written for the evaporation layer and the glass cover of the interfacial evaporation structure. For the evaporation layer, a portion of the absorbed solar radiation energy is transferred to the glass cover through evaporation, convection, and radiation, a portion is lost to the sidewalls and the main body of water, and the remaining portion heats the evaporation layer. The corresponding energy balance equation is expressed as follows:

$$\rho_e c_{p,e} L_e A_e \cdot \frac{dT_e}{dt} = \alpha A_e \cdot q_{solar} - A_e q_{evap} - A_e q_{conv} - A_e q_{rad} - A_{side} q_{side,e} - A_e q_{bottom}, \quad (1)$$

where ρ_e is the evaporation layer density, $c_{p,e}$ is the specific heat of the evaporation layer, L_e is the evaporation layer thickness, T_e is the evaporation temperature, t is time, and α is the solar absorptivity of the evaporation layer.

For the glass cover, the heat transferred from the evaporation surface to the glass plate is equal to the heat transfer between the glass cover plate and the external environment and the temperature rise of the glass cover, the energy balance equation can be expressed as,

$$c_{p,g}\rho_g L_g A_g \frac{dT_g}{dt} = \alpha_g A_g \cdot q_{solar} + A_g q_{conv} + A_g q_{rad} + A_g q_{evap} - A_g q_{rad,g-a} - A_g q_{conv,g-a} - A_{side} q_{side,inside}, \quad (2)$$

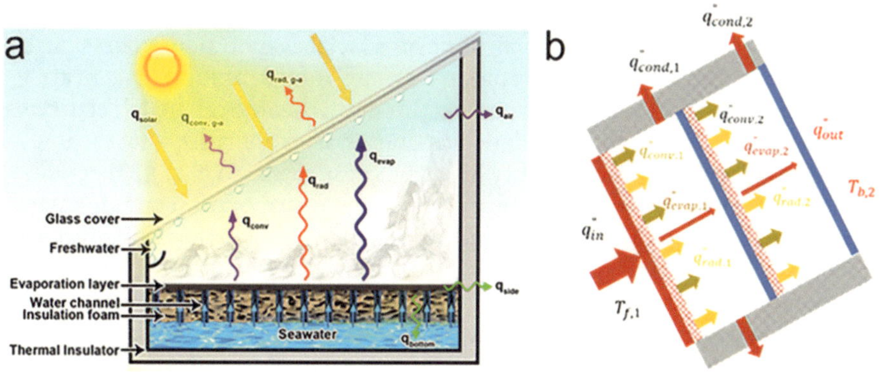

Figure 1.
Schematic diagram of physical model of solar still. (a) Heat transfer diagram of interfacial evaporation solar still, and (b) Heat transfer diagram of multistage condensation system.

where $c_{p,g}$ is the specific heat of the glass cover, ρ_g is the density of the glass cover, L_g is the thickness of glass cover, T_g is the glass cover temperature, $q_{rad,g-a}$ is the radiation heat transfer flux from evaporation layer to the sky, $q_{conv,g-a}$ is the convection heat transfer flux from glass cover to ambient air, and $q_{side,inside}$ is the heat loss through the side wall by internal wet air.

By solving Eqs. (1) and (2), the evaporation surface temperature, glass cover temperature as a function of time could be obtained. Thus, the water production rate $\dot{m}$ can be calculated,

$$\dot{m} = h_{evap,e-g}\frac{T_e - T_g}{h_{fg}}, \tag{3}$$

where h_{fg} is latent heat of vaporization on the T_g.

In addition to solar stills, heat transfer models can also be established and analyzed for multistage condensation systems. For each stage of the multistage condensation structure, the energy balance equation is expressed as

$$Q''_{\text{in}} = Q''_{\text{conv}} + Q''_{rad} + Q''_{\text{evap}} + Q''_{\text{conduction}} \tag{4}$$

where Q''_{in} is the heat flux from the condensation surface to the multistage condensation structure, Q''_{conv} is the convection heat transfer flux from this stage to the next stage, Q''_{rad} is a radiation heat transfer from this stage to next stage, Q''_{evap} is the evaporation heat transfer flux from this stage to the next stage, and $Q''_{\text{conduction}}$ is the conduction heat transfer from this stage to the next stage.

According to the theoretical model, by numerically solving Eqs. (1) and (2), the temperature distribution of the evaporation surface and the glass plate of the solar still can be obtained, and the fresh water production rate can be obtained at the same time. In order to accurately calculate the temperature of each surface of the multistage condensation structure, an initial value of Q_{in} needs to be given, and the temperature distribution of the multistage condensation structure can be calculated by Eq. (4) and correct the temperature and Q''_{in} of the condensation surface through iterative methods to obtain the temperature distribution of each position of multistage condensation structure. According to the analysis, improving the water production efficiency can be achieved by changing A_e, q_{conv}, q_{rad}, q_{side}, and $q_{conv,g-a}$. Thus, the main factors affecting the efficiency of the solar still are the adiabatic performance of the solar still, the emissivity of the light-absorbing material, the external convective heat transfer term, and the internal convective heat transfer term.

Based on the aforementioned model, the interfacial evaporation process can be accurately described. Shi et al. [28] analyzed the advantages of interfacial heating compared to traditional volumetric heating through a theoretical model, clarifying the precise calculation method for dark evaporation rates and analyzing the impact of ambient temperature and humidity on evaporation rates. Xu et al. [33] conducted a simulation to analyze the temperature variations and evaporation rates in each stage of a multistage condensation system, performing parameter analysis, and proposed a multistage condensation system with higher freshwater yield. Through transient calculations of the model, Luo et al. [30] performed a detailed analysis of the heat and mass transfer processes in interfacial evaporation solar stills and, based on the parameter analysis results, proposed optimization methods suitable for interfacial evaporation stills, significantly enhancing the freshwater yield of solar stills. Chen et al. [35] also applied this model to the analysis of multistage interfacial evaporation systems, conducting detailed parameter analysis and experimental validation, and calculated the theoretical maximum efficiency of multistage interfacial evaporation systems, providing theoretical guidance for the development of multistage interfacial evaporation systems.

3. Optimization method for interfacial evaporation solar stills

3.1 Increasing evaporation area

In recent years, through material innovation and thermal management optimization, interfacial evaporation structures have achieved over 80% evaporation efficiency under an illumination intensity of 1 kW/h. Theoretical analysis and other literature studies [36] have shown that to further enhance the evaporation rate, increasing the surface area of the evaporating interface can strengthen the evaporation process, as illustrated in **Figure 2**.

By increasing the evaporation area while maintaining the same illumination area, the intensity of illumination per unit area decreases, resulting in a high evaporation rate. Li et al. [37] discovered that in an open evaporation environment, utilizing the dark evaporation rate and temperature gradient of the photothermal evaporation surface can promote evaporation from other surfaces, achieving a higher evaporation rate of up to 4.1 kg·m^{-2}·h^{-1}, which exceeds the evaporation limit calculated from the enthalpy of evaporation. To explain this phenomenon, the authors proposed that

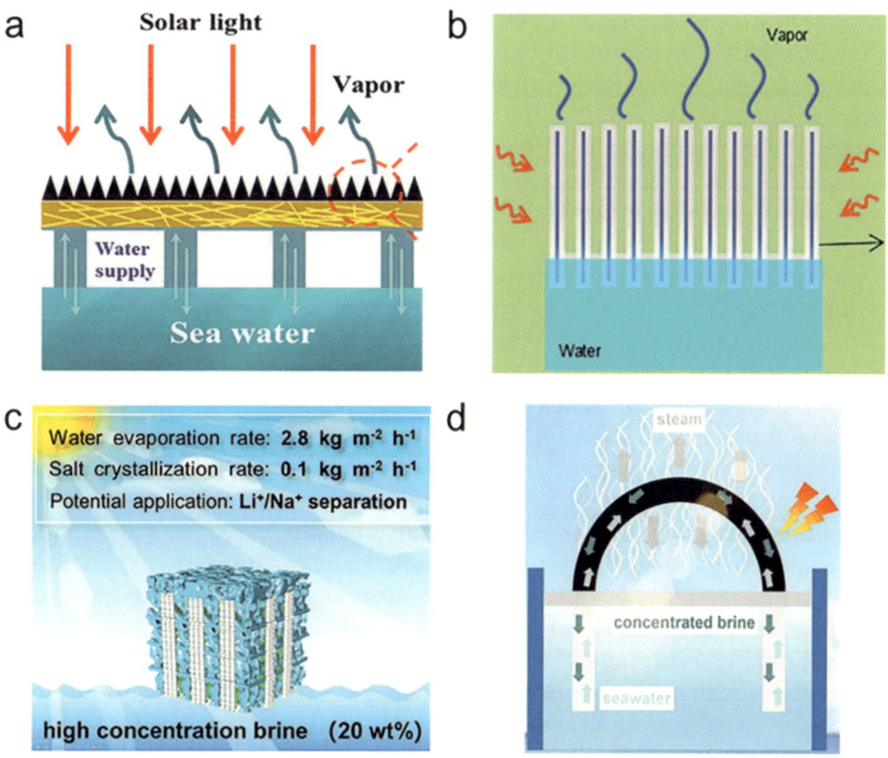

Figure 2.
The influence of evaporation area on evaporation performance [37–40].

the evaporation process might cause the evaporation surface temperature to drop below the ambient temperature, thereby absorbing energy from the environment to facilitate evaporation. Consequently, more studies have been developing 3D evaporators to enhance evaporation rates. However, Luo et al. [5] indicate that increasing the evaporation temperature through concentration of light can improve the evaporation rate by increasing the illumination intensity per unit area, thus enhancing the evaporation efficiency, which is opposite to the theory that increasing the evaporation area can improve the evaporation rate.

Based on these two differing viewpoints, using the heat transfer model mentioned above as the basis, Luo et al. [28] conducted theoretical analysis to assess the impact of changing the evaporation area on evaporation performance by studying the variation in evaporation efficiency with increased illumination intensity per unit area. Additionally, they evaluated whether changing the evaporation area in a solar still would affect desalination performance under different humidity levels. The study found that when ϕ = 20% and 60%, as the illumination intensity increases, the evaporation efficiency initially decreases and then increases, eventually stabilizing. Under a constant energy input of 1 $kW{\cdot}m^{-2}$, the evaporation efficiency curve with illumination intensity per unit area lower than 1 $kW{\cdot}m^{-2}$ represents the increased evaporation area. The results show that increasing the evaporation area in an open environment continuously raises the evaporation efficiency, even approaching infinity. This does not violate the law of energy conservation, as evaporation is a continuous process driven by the vapor pressure difference between the evaporation surface and the environment. When there is no illumination intensity, the evaporation material absorbs environmental energy to proceed with evaporation. Therefore, increasing the evaporation area in an open environment can significantly enhance evaporation efficiency. However, in a closed environment (ϕ = 100%), increasing the evaporation area gradually reduces the evaporation efficiency, approaching zero. This is because when the relative humidity nears 100%, the vapor pressure difference between the water vapor and the environment significantly decreases, causing nonsolar-driven evaporation to gradually disappear. The vapor pressure around the evaporation material, which gradually heats up under illumination, increases, promoting the evaporation process. Thus, in an environment where relative humidity nears 100%, increasing the evaporation area does not effectively improve the evaporation efficiency of the material.

In solar stills, the freshwater production rate of interfacial evaporation structures with different evaporation surface areas and volumetric heating was calculated [28], as shown in **Figure 3b**. When the illumination area is 0.25 m^2 and the evaporation area increases from 0.25 to 3 m^2, the freshwater production rate of the distiller remains approximately 4.3 $kg{\cdot}m^{-2}$, not only failing to increase but slightly decreasing. The study revealed that after increasing the evaporation area, the energy received per unit area decreases under the same illumination intensity, leading to a lower evaporation temperature and evaporation rate, thereby reducing freshwater yield. In an open environment, increasing the evaporation area raises the steam production rate, but the thermal quality of the water vapor is low, preventing effective condensation and collection. Previous studies [41] on solar stills have shown that increasing the evaporation area can increase water production by 10–20%. This could be due to the fact that when the evaporation area is increased by designing a stepped evaporation surface, the equivalent illumination area increases, leading to increased solar flux and evaporation rate. Furthermore, many studies [37–40] have optimized the evaporation surface using porous media, altering the heat capacity of the water to be evaporated,

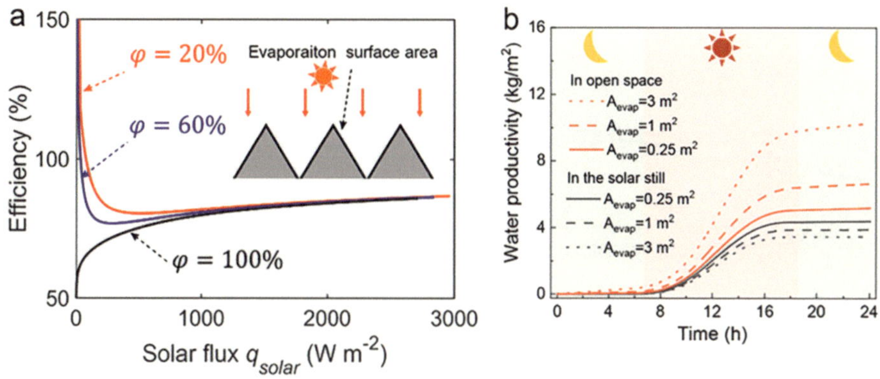

Figure 3.
The influence of interfacial evaporation structure evaporation area on evaporation performance [28]. (a) The effect of solar heat flux on evaporation efficiency under different humidity levels; and (b) The effect of increasing the evaporation surface area inside a solar still on freshwater production.

rather than merely increasing the evaporation area. Therefore, in solar stills, increasing the evaporation area of the evaporation material without changing the illumination area does not effectively improve freshwater production and may even reduce it. Based on the above research results, it can be observed that in an open environment or under conditions of enhanced evaporation, increasing the evaporation area can effectively improve the evaporation rate. However, in solar stills, increasing the evaporation area does not effectively increase the freshwater yield.

3.2 Strengthening heat transfer process

To enhance the freshwater yield of solar stills, strengthening the heat transfer process within the solar still is one of the measures that can improve the desalination performance, according to parameter analysis from theoretical models and insights from traditional distiller optimization methods. In traditional solar stills, the freshwater yield can be optimized by lowering the temperature of the glass cover through forced convection, which enhances the condensation process and increases the steam condensation rate, thereby boosting freshwater production. Additionally, intensifying the convective heat transfer process within the distiller can promote water vapor condensation, reduce the relative humidity inside the distiller, and effectively increase the condensation rate. Consequently, many studies [32, 33] have analyzed the impact of enhanced heat transfer methods on the performance of interfacial evaporation solar stills, as shown in **Figure 4**. In traditional solar still designs [32, 33], external forced convection measures such as water film cooling are typically used to optimize freshwater output. Yang et al. [20] used water film cooling to increase the freshwater output of a volumetric heating solar still by approximately 10%. However, whether this method is suitable for interfacial evaporation structures requires further calculation and analysis.

Luo et al. [30] conducted theoretical analysis and experimental research on the impact of heat transfer enhancement on interfacial evaporation solar stills. They studied the effects of convection enhancement on the outer surface of the still's glass cover and convection enhancement inside the still on freshwater yield, exploring methods to improve the performance of interfacial evaporation stills. The study found that for interfacial heating structures, the thin water layer on the evaporation surface and its low

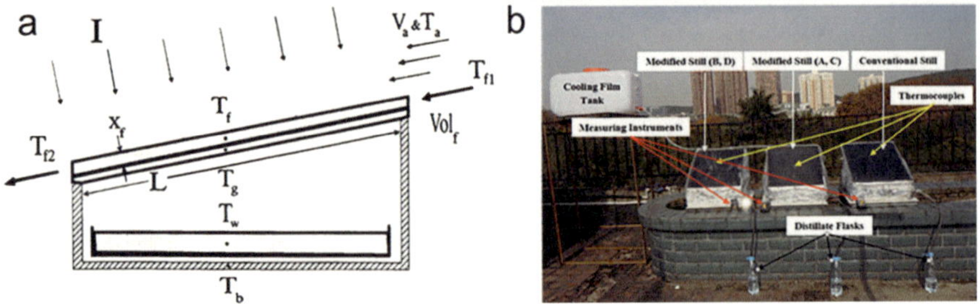

Figure 4.
The influence of interface evaporation structure evaporation area on evaporation performance [32, 33].

thermal capacity mean that its temperature is significantly affected by changes in the glass cover temperature. When the convective heat transfer coefficient is low, increasing it lowers the temperature of the glass cover, enhancing the condensation process and increasing the amount of condensation. When the external convective heat transfer process of the glass cover is enhanced, the glass cover temperature significantly decreases, which reduces the evaporation surface temperature and evaporation rate, thereby inhibiting the evaporation process and decreasing the freshwater yield. By using a fan to disturb the air inside the solar still, the heat and mass transfer process between the evaporation surface and the glass cover is enhanced, making it easier for water vapor to reach the inner surface of the glass cover for condensation. Simultaneously, the evaporation process intensifies, leading to an increased evaporation rate on the evaporation surface, thereby improving the freshwater productivity in the solar still. Therefore, enhancing the internal convective heat transfer process of the solar still is an effective optimization method suitable for promotion and application, while water film cooling is not suitable for performance enhancement of interfacial evaporation solar stills (**Figure 5**).

3.3 Other methods

One optimization measure to enhance the evaporation rate is to reduce the radiative heat loss of evaporation materials by lowering their emissivity. Therefore, seeking photothermal conversion materials with high absorptivity and low emissivity can improve the evaporation rate [33]. According to Kirchhoff's law, the absorptivity of a medium in thermal equilibrium should equal its emissivity. However, the high absorptivity and low emissivity evaporation materials referred to here are those with high absorptivity in the ultraviolet–visible-near-infrared spectrum and low emissivity in the far-infrared spectrum. This allows the evaporation material to maximize solar energy absorption while minimizing radiative heat loss [33].

Nevertheless, the impact of the emissivity variation of light-absorbing materials on the water production rate of solar stills requires further research to evaluate the cost–benefit ratio of this optimization measure. Some researches [30] found that the effect of thermal radiation loss on the water production rate is minimal, and reducing the emissivity of evaporation materials can be disregarded when optimizing the solar still (**Figure 6**).

In previous studies, the inclusion of an insulation layer within the interfacial evaporation structure meant that additional insulation was not required in evaporation and seawater desalination experiments, thereby reducing construction costs. However, Luo et al. [30] discovered that for practical stills, where the evaporation

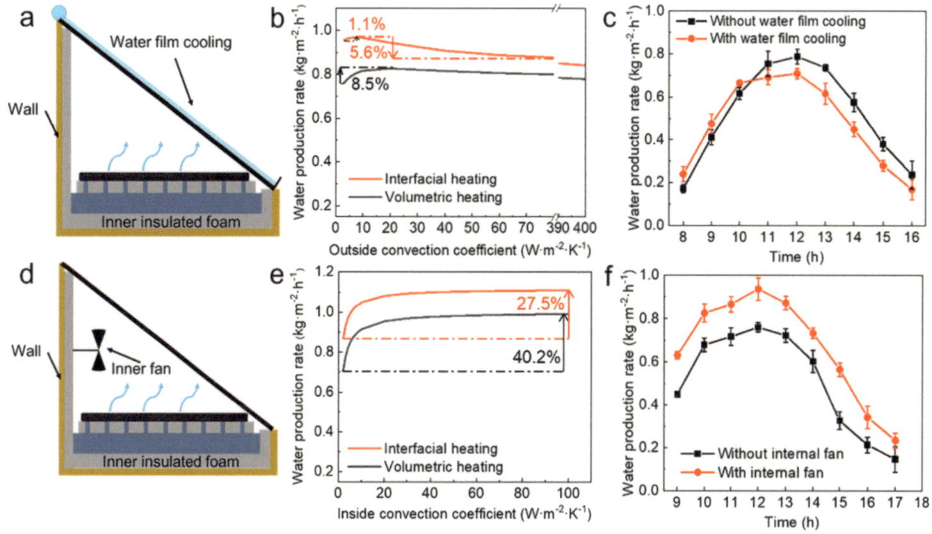

Figure 5.
Performance of interfacial evaporation solar still under different convective heat transfer coefficient [30]. (a) Convective heat transfer coefficient between evaporation surface and air in the solar still; (b) Effect of internal fan inside the solar still on water production rate; (c) Effect of internal fan inside the solar still on temperature of different locations; and (d) Effect of internal fan inside the solar still on water production rate of a day under one sun.

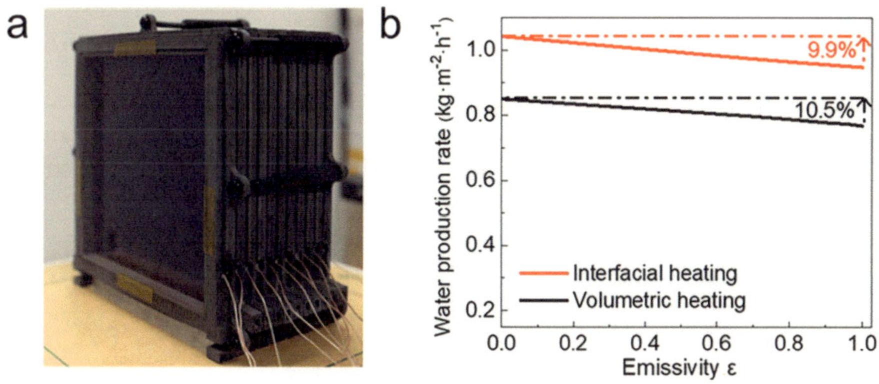

Figure 6.
The influence of surface emissivity variation on evaporation performance. (a) Selective light-absorbing materials [33]; and (b) Theoretical analysis of the effect of surface emissivity changes on freshwater production in solar stills [30].

interface contacts the walls, there is heat leakage through the sidewalls. To reduce heat loss from the evaporation surface to the water body, insulation of the walls can be enhanced. Experimental results indicate that insulating the inner walls can more effectively reduce heat loss.

3.4 Waste heat recovery

According to literature analysis, another effective method to increase freshwater yield is waste heat recovery. In the design of solar distillers, the latent heat released during the condensation of water vapor on the glass cover, which turns into freshwater, is typically lost to the ambient air without being effectively reused. Inspired by traditional multieffect condensation solar stills [32, 42], researchers [43] have designed a multi-stage condensation system suitable for thin-film evaporation in interfacial evaporation structures. This system uses hydrophobic membranes like polytetrafluoroethylene as separating membranes between seawater and freshwater layers, allowing water vapor to pass through the membrane and condense in the freshwater layer. The latent heat from condensation is then transferred to the next seawater layer, preventing seawater contamination of freshwater and achieving multistage condensation.

By increasing the number of condensation stages, the latent heat from condensation can be utilized more effectively. When the condensation technology is improved to 10 stages, the condensation yield is significantly enhanced. Asinari et al. [34] demonstrate a ten-stage condensation system to achieve a freshwater production rate of 2.479 $kg{\cdot}m^{-2}{\cdot}h^{-1}$, while Wang et al. [44] demonstrated a flexible eight-stage condensation system to achieve a freshwater production rate of 3.61 $kg{\cdot}m^{-2}{\cdot}h^{-1}$ under a single light intensity. Xu et al. [33] proposed a ten-stage condensation structure that achieved a steam production rate of 4.85 $kg{\cdot}m^{-2}{\cdot}h^{-1}$ under a single light intensity, greatly increasing the freshwater yield in nonconcentrated light environments.

In the study of solar seawater desalination systems, multistage condensation structures are effective waste heat recovery devices. The design of distillers should fully consider the application of multistage condensation structures. By utilizing waste heat recovery, the freshwater yield of the seawater desalination system can be further

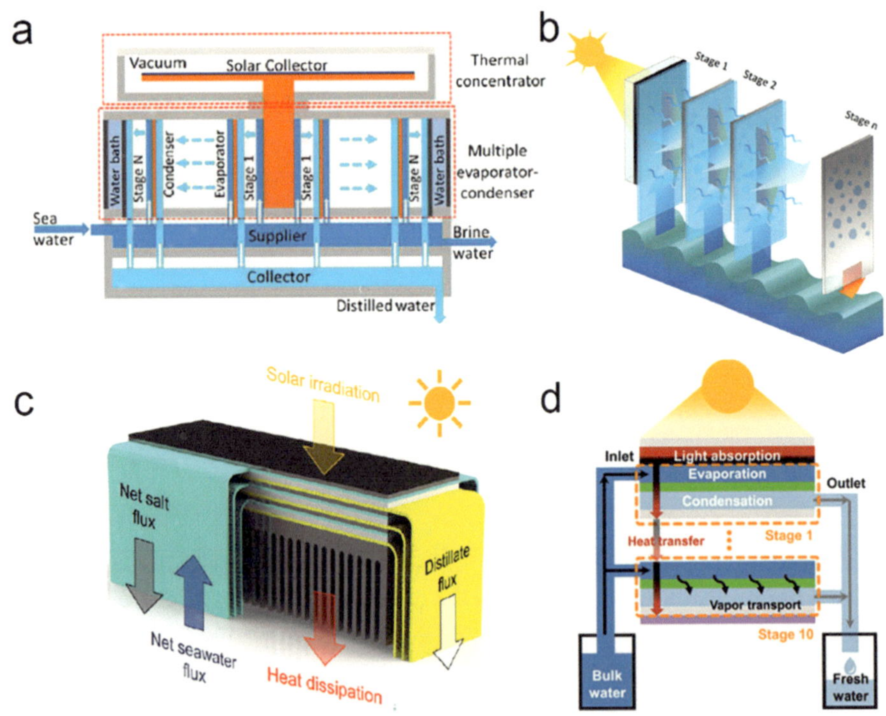

Figure 7.
Research on multistage condensation system: (a) Circular multistage condensation system [45]; (b) Ten-level vertical multistage condensation system [33]; (c) Horizontal multistage condensation system [34]; and (d) Multistage condensation system for optimizing evaporative surfaces [46].

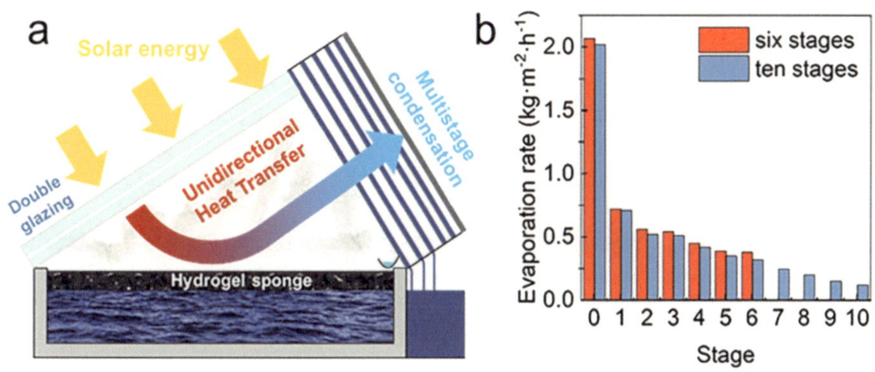

Figure 8.
Multistage condensation-unidirectional heat transfer solar still [47]. (a) Schematic diagram of a multistage condensation-unidirectional heat transfer solar still, and (b) The water collection rate of the UHTSS with multistage condensation.

improved, achieving high freshwater production output at low cost with no additional energy input, relying solely on solar energy (**Figure 7**).

To further study multistage interfacial distillation systems, Chen et al. [35] conducted a detailed analysis of energy flow during the desalination process of multistage interfacial stills. They calculated the upper limit of energy conversion efficiency, identified the primary mechanisms of energy loss, and explored methods to enhance energy efficiency. This research provides valuable guidance for the design of materials and systems aimed at improving the energy efficiency of multistage interfacial seawater desalination. To further enhance the freshwater yield of solar stills employing interfacial evaporation structures, Luo et al. [47] redesigned the heat transfer channels within the solar distiller and proposed an efficient unidirectional heat transfer solar still. By separating the light transmission and condensation functions within the glass cover and optimizing them individually, the overall heat transfer process was improved. Additionally, to further boost the distiller's freshwater yield, this distiller was combined with custom-made hydrogel sponges and a multistage condensation structure to create a multistage condensation-unidirectional heat transfer solar still. This integrated design achieved efficient and continuous freshwater production, yielding systematic and comprehensive experimental data and results. Furthermore, the large-scale expandable hydrogel sponges, scalable multistage condensation structures, and the composite design of large solar distillers laid a solid foundation for the industrial-scale application of efficient solar seawater desalination systems (**Figure 8**).

4. Conclusion

In summary, this chapter discusses the performance optimization methods for interfacial evaporation solar stills in recent studies. It has been found that while increasing the evaporation area can improve the evaporation rate, it does not enhance the freshwater yield within the still. Forced convection heat transfer inside the still, compared to forced convection heat transfer on the outer surface of the glass cover, is more effective in increasing the freshwater yield. For interfacial evaporation solar stills, due to the risk of heat leakage from the evaporation surface along the sides to larger water bodies, installing insulation material on the inner walls of the solar still can effectively increase the freshwater yield. Finally, it has been found that utilizing waste heat is an effective method to significantly increase freshwater yield, potentially achieving multiple-fold growth in output. Additionally, combining waste heat utilization devices with optimized interfacial evaporation stills can enable efficient and continuous freshwater production, demonstrating potential for large-scale and scalable applications.

This chapter investigates the optimization of interfacial evaporation solar stills from the perspective of heat and mass transfer. Future research can be further explored from the following angles. Firstly, the theoretical mechanisms of photothermal conversion materials still require in-depth investigation. Particularly for materials such as hydrogels that can achieve ultra-high evaporation rates, further exploration is needed to provide more reasonable explanations from both theoretical and experimental perspectives and to offer guidance for other photothermal conversion materials. Secondly, to enhance the freshwater yield of solar stills in natural environments, research should focus on maintaining high freshwater output under varying conditions, optimizing freshwater yield in low-light and no-light environments. Finally, to further increase the amount of condensed water collected, efforts should be directed toward improving condensation collection techniques without affecting the evaporation rate, thereby enhancing freshwater productivity.

Author details

Xiao Luo[1,2] and Xiaokun Gu[1*]

1 Institute of Thermophysics, School of Mechanical Engineering, Shanghai Jiao Tong University, Shanghai, China

2 Shenergy Co. Ltd, Shanghai, China

*Address all correspondence to: xiaokun.gu@sjtu.edu.cn

References

[1] Qiblawey H, Banat F. Solar thermal desalination technologies. Desalination. 2008;**220**:633-644

[2] Thirugnanasambandam M, Iniyan S, Goic R. A review of solar thermal technologies. Renewable and Sustainable Energy Reviews. 2010;**14**:312-322

[3] Ghasemi H, Ni G, Marconnet AM, Loomis J, Yerci S, Miljkovic N, et al. Solar steam generation by heat localization. Nature Communications. 2014;**5**:4449

[4] Tao P, Ni G, Song C, Shang W, Wu J, Zhu J, et al. Solar-driven interfacial evaporation. Nature Energy. 2018;**3**:1031-1041

[5] Pang Y, Zhang J, Ma R, Qu Z, Lee E, Luo T. Solar–thermal water evaporation: A review. ACS Energy Letters. 2020;**5**:437-456

[6] Zhu L, Gao M, Peh CKN, Ho GW. Recent progress in solar-driven interfacial water evaporation: Advanced designs and applications. Nano Energy. 2019;**57**:507-518

[7] Chen C, Kuang Y, Hu L. Challenges and opportunities for solar evaporation. Joule. 2019;**3**:683-718

[8] Liu S, Huang C, Luo X, Guo C. Performance optimization of bi-layer solar steam generation system through tuning porosity of bottom layer. Applied Energy. 2019;**239**:504-513

[9] Wang X, He Y, Liu X, Cheng G, Zhu J. Solar steam generation through bio-inspired interface heating of broadband-absorbing plasmonic membranes. Applied Energy. 2017;**195**:414-425

[10] Ito Y, Tanabe Y, Han J, Fujita T, Tanigaki K, Chen M. Multifunctional porous graphene for gigh-efficiency steam generation by heat localization. Advanced Materials. 2015;**27**:4302-4307

[11] Zhou L, Tan Y, Ji D, Zhu B, Zhang P, Xu J, et al. Self-assembly of highly efficient, broadband plasmonic absorbers for solar steam generation. Science Advances. 2016;**2**:e1501227

[12] Zhou L, Tan Y, Wang J, Xu W, Yuan Y, Cai W, et al. 3D self-assembly of aluminium nanoparticles for plasmon-enhanced solar desalination. Nature Photonics. 2016;**10**:393-398

[13] Li X, Xu W, Tang M, Zhou L, Zhu B, Zhu S, et al. Graphene oxide-based efficient and scalable solar desalination under one sun with a confined 2D water path. Proceedings of the National Academy of Sciences of the United States of America. 2016;**113**:13953-13958

[14] Wang Y, Zhang L, Wang P. Self-floating carbon nanotube membrane on macroporous silica substrate for highly efficient solar-driven interfacial water evaporation. ACS Sustainable Chemistry & Engineering. 2016;**4**:1223-1230

[15] Li Y, Gao T, Yang Z, Chen C, Luo W, Song J, et al. 3D-printed, all-in-one evaporator for high-efficiency solar steam generation under 1 sun illumination. Advanced Materials. 2017;**29**:1700981

[16] Tian Y, Yang H, Wu S, Yan J, Cen K, Luo T, et al. Beyond lotus: Plasma nanostructuring enables efficient energy and water conversion and use. Nano Energy. 2019;**66**:104125

[17] Tiwari AK, Tiwari GN. Effect of water depths on heat and mass transfer in a passive solar still: in summer climatic condition. Desalination. 2006;**195**:78-94

[18] Tarawneh M. Effect of water depth on the performance evaluation of solar still. Jordan Journal of Mechanical and Industrial Engineering. 2007;**1**(1):23-29

[19] Tanaka H. Experimental study of a basin type solar still with internal and external reflectors in winter. Desalination. 2009;**249**:130-134

[20] Peng G, Sharshir S W, Hu Z, et al. A Compact Flat Solar Still With High Performance. International Journal of Heat and Mass Transfer. 2021;**179**:121657

[21] Panchal H, Hishan SS, Rahim R, Sadasivuni KK. Solar still with evacuated tubes and calcium stones to enhance the yield: An experimental investigation. Process Safety and Environmental Protection. 2020;**142**:150-155

[22] Nijmeh S, Odeh S, Akash B. Experimental and theoretical study of a single-basin solar still in Jordan. International Communications in Heat and Mass Transfer. 2005;**32**:565-572

[23] Dunkle RV. Solar Water Distillation, the Roof Type Still and a Multiple Effect Diffusion Still. International Developments in Heat Transfer. ASME, Proceeding of International Heat Transfer. Part V. University of Colorado. 1961:**895**

[24] Okujagu CU, Adjepong SK. Performance of a simple flat plate solar collector at an equatorial location. Solar and Wind Technology. 1989;**6**:283-289

[25] Xiao G, Wang X, Ni M, Wang F, Zhu W, Luo Z, et al. A review on solar stills for brine desalination. Applied Energy. 2013;**103**:642-652

[26] Panchal HN. Life cycle cost analysis of a double-effect solar still. International Journal of Ambient Energy. 2017;**38**:395-399

[27] Mohan I, Yadav S, Panchal H, Brahmbhatt S. A review on solar still: A simple desalination technology to obtain potable water. International Journal of Ambient Energy. 2019;**40**:335-342

[28] Luo X, Shi J, Zhao C, Luo Z, Gu X, Bao H. The energy efficiency of interfacial solar desalination. Applied Energy. 2021;**302**:117581

[29] Sharshir SW, Peng G, Wu L, Essa FA, Kabeel AE, Yang N. The effects of graphite nanoparticles, phase change material, and film cooling on the solar still performance. Applied Energy. 2016;**191**:358-366

[30] Luo X, Gu X, Shi J, Zhao C, Bao H. Thermal design strategy for enhanced freshwater harvesting with interfacial evaporation. Applied Thermal Engineering. 2022;**216**:119104

[31] Park C-D, Lim B-J, Chung K-Y, Lee S-S, Kim Y-M. Experimental evaluation of hybrid solar still using waste heat. Desalination. 2016;**379**:1-6

[32] Tanaka H, Nosoko T, Nagata T. A highly productive basin-type-multiple-effect coupled solar still. Desalination. 2000;**130**:279-293

[33] Zhenyuan X, Zhang L, Zhao L, Li B, Bhatia B, Chenxi Wang KL, et al. Ultrahigh-efficiency desalination via a thermally-localized multistage solar still. Energy and Environmental. 2020;**13**:830-839

[34] Chiavazzo E, Morciano M, Viglino F, Fasano M, Asinari P. Passive solar high-yield seawater desalination by modular and low-cost distillation. Nature Sustainability. 2018;**1**:763-772

[35] Chen Y, Luo X, Gu X, Bao H. The energy efficiency of multistage interfacial desalination: Comprehensive

analysis and further improvement strategy. Desalination. 2023;**568**:117000

[36] Wang Y, Wu X, Yang X, Owens G, Xu H. Reversing heat conduction loss: Extracting energy from bulk water to enhance solar steam generation. Nano Energy. 2020;**78**:105269

[37] Li X, Li J, Lu J, Xu N, Chen C, Min X, et al. Enhancement of interfacial solar vapor generation by environmental energy. Joule. 2018;**2**:1331-1338

[38] Zhao Y, You D, Chen Y, Pan Q, Su Z, Yang W. Highly efficient 3D evaporator for interfacial solar steam generation and wastewater treatment. Energy Technology. 2024;**12**:2300586

[39] Chen Z, Li Q, Chen X. Porous graphene/polyimide membrane with a three-dimensional architecture for rapid and efficient solar desalination via interfacial evaporation. ACS Sustainable Chemistry and Engineering. 2020;**8**:13850-13858

[40] Zhang Y, Zhong Q, Huang Q, Hu M, He F, Li Y, et al. Surface engineering of 3D solar evaporator for uncompromising water evaporation and salt production toward high concentration brine. Advanced Functional Materials. 2024;**2408554**

[41] Peng G, Sharshir SW. Progress and performance of multi-stage solar still–a review. Desalination. 2023;**565**:116829

[42] Tanaka H, Iishi K. Experimental study of a vertical single-effect diffusion solar still coupled with a tilted wick still. Desalination. 2017;**402**:19-24

[43] Wang W, Shi Y, Zhang C, Hong S, Shi L, Chang J, et al. Simultaneous production of fresh water and electricity via multistage solar photovoltaic membrane distillation. Nature Communications. 2019;**10**:3012

[44] Wang W, Shi Y, Zhang C, Li R, Wu M, Zhuo S, et al. Solar seawater distillation by flexible and fully passive multistage membrane distillation. Nano Letters. 2021;**21**:5068-5074

[45] Huang L, Jiang H, Wang Y, Ouyang Z, Wang W, Yang B, et al. Enhanced water yield of solar desalination by thermal concentrated multistage distiller. Desalination. 2020;**477**:114260

[46] Liu S, Li S, Lin M. Understanding interfacial properties for enhanced solar evaporation devices: From geometrical to physical interfaces. ACS Energy Letters. 2023;**8**:1680-1687

[47] Luo X, Jiao L, Guo Y, Bao H, Zhao C, Gu X. Ultrahigh freshwater production achieved by unidirectional heat transfer interfacial evaporation solar still integrated with waste heat recovery. Energy Conversion and Management. 2024;**304**:118226

Chapter 3

Breakthroughs and Prospects: The Development Path of Solar Thermal Seawater Desalination Technology

Huijin Xu and Hao Luo

Abstract

Amid the escalating severity of water scarcity, seawater desalination has emerged as a viable solution. Traditional seawater desalination technologies are fraught with numerous drawbacks, including high energy consumption, exorbitant costs, intricate structures, and adverse environmental impacts. In contrast, solar-powered seawater desalination technology has garnered significant attention due to its economic viability and environmental sustainability. This article undertakes a comprehensive study of the solar seawater desalination system, encompassing both direct and indirect methodologies, and delves into the strengths and weaknesses of diverse technologies, such as solar distillation, humidification-dehumidification, solar chimney, reverse osmosis, multi-effect distillation, and multi-stage flash evaporation. The direct approach harnesses solar energy to directly desalinate seawater, whereas the indirect method transforms solar energy into other energy forms for the purpose of seawater desalination. Simultaneously, an in-depth analysis of the benefits of solar-powered seawater desalination technology is conducted. Its economic merits include minimal infrastructure requirements and ease of local design, installation, and maintenance, albeit with a limited water production capacity per unit area. The costs associated with different technologies exhibit substantial variations and are subject to the influence of multiple factors. Future development trajectories are centered around enhancing efficiency, reducing costs, and achieving large-scale applications, exemplified by the exploration of hybrid rare-earth-driven desalination, forward osmosis, and dew point evaporation technologies, as well as the integration of advanced materials, intelligent control systems, and energy storage technologies.

Keywords: solar-powered seawater desalination, solar photovoltaic-reverse osmosis, multi-effect distillation, multi-stage flash evaporation, dew point evaporates

1. Introduction

Water, as the source of life, is crucial for all life forms on Earth. On the vast Earth, approximately 97.5% of water is in a salty state, meaning that only 2.5% of fresh water can meet human usage needs. For modern society, it plays a vital role in food security, health, housing, energy production, industry, and economic growth. However, with the growth of the global population and economic development, the shortage of water resources has become a major challenge facing the world. Problems such as industrial wastewater discharge, over-exploitation, and environmental pollution have led to an increasingly acute contradiction between supply and demand of freshwater resources. According to the United Nations report [1], the United Nations World Water Development Report points out that global freshwater usage is growing at a rate of just under 1% per year, driven by economic and social development and changes in consumption patterns, including diet. Although agricultural water uses accounts for about 70% of freshwater consumption, industrial (about 20%) and domestic water (about 10%) are the main areas where freshwater demand is increasing [2]. A recent study estimates that nearly 5 billion people live in areas where water security may be threatened. To meet the needs of the global population, there is an urgent need to seek alternative sources of freshwater beyond natural resources.

To address this situation, distillation processes have been used for decades. More specifically, desalination processes are being employed, where saline water is treated to obtain freshwater. Seawater desalination is essentially a process that separates freshwater from brackish or saline water. Some energy is required to run this process. Thermal energy is widely used to operate seawater desalination plants, which is derived from fossil fuels. It is apparent that these processes require a significant amount of energy, and due to high costs, complex structures, a large CO_2 footprint, and global warming effects, they are not economically and ecologically viable. Moreover, it is not a permanent energy source. While many seawater desalination plants operate on fossil fuels and produce millions of liters of fresh water from saline or brackish water at the cost of climate change, leading to global warming, energy sustainability is one of today's most pressing socio-environmental issues. Up to 80% of the world's energy comes from fossil fuels, which has devastating effects on the environment. In most cases, the ever-growing global energy demand is met by increasing fossil fuel consumption, resulting in acidification, global warming, air pollution, and land degradation. The capacity and efficiency of energy infrastructure, along with the broad availability of affordable energy (especially electricity), are crucial for socio-economic equality [3].

The energy sector is undergoing a crucial transformation from traditional fossil fuels to renewable energy. Renewable energy, with its remarkable advantages such as sustainability and low environmental impact, is gradually becoming the core force in responding to the challenges of the energy crisis and climate change, bringing brand-new development opportunities and directions of transformation for numerous industries. The seawater desalination process, as a key technological field concerning the security and sustainable utilization of global water resources, is also in a transitional stage to meet the growing water demand of human beings. In recent years, various methods for seawater desalination have been developed and tested all over the world. Most of the development projects are concentrated in the Middle Eastern Arab countries, African countries, China, Australia, India, and other regions where freshwater resources are not abundant [4]. Among numerous seawater desalination technologies, solar energy, as one of the most abundant and widely distributed forms of renewable energy, has demonstrated huge application potential and has given birth to the research direction of solar-powered seawater desalination technology, which is a highly promising field. The solar-powered seawater desalination technology aims to utilize solar energy, a clean energy source, and convert seawater into directly usable fresh water through innovative technical means, thus providing a green, efficient, and economically viable solution to the water shortage problems in coastal areas and regions with scarce freshwater resources. Among them, the solar still, as an important part of the solar-powered seawater desalination technology system, relies on its characteristics such as relatively simple structure, convenient operation, and high reliability. In **Figure 1**, the compound parabolic concentrator tubular solar still. Through the concentrating effect of the compound parabolic concentrator (CPC), more solar energy can be gathered into seawater, increasing the temperature of the seawater and accelerating the evaporation process. As a result, the conversion efficiency from solar energy to thermal energy is improved and solar energy resources are utilized more effectively. Its structural design enables the processes of seawater evaporation and water vapor condensation to be carried out in a relatively stable environment. The design of the concentric tubes is conducive to the transmission and condensation of water vapor, which can continuously and stably produce fresh water. In the future, more solar desalination technologies are expected to be more widely applied.

This book conducts research and analysis on solar desalination systems (including both indirect and direct methods) and explores novel and traditional desalination technologies. It discusses the advantages and disadvantages of various solar desalination methods based on the economic efficiency and technical data of most plants. This article is divided into six parts: Chapter One is the introduction, Chapter Two introduces the classification of solar desalination systems, Chapter Three analyzes the technical advantages of solar desalination, Chapter Four discusses the degree of commercialization of solar desalination technologies, Chapter Five provides future prospects, and Chapter Six is the summary.

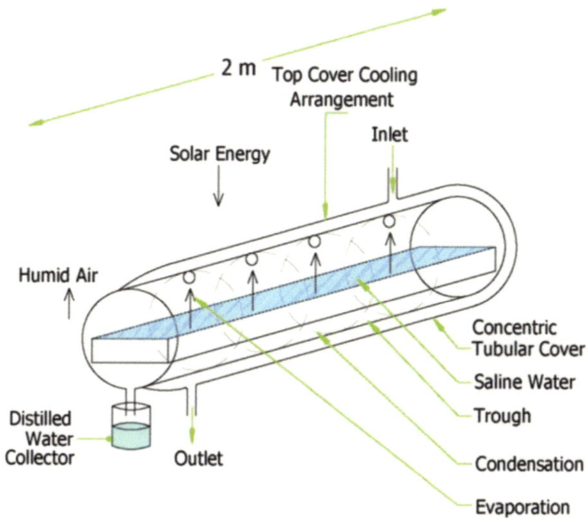

Figure 1.
Compound parabolic concentrator tubular solar still [5].

2. Solar desalination

2.1 Working principles of desalination

The basic principle of seawater desalination is to remove salt and other impurities from seawater, converting it into fresh water suitable for drinking or industrial use. This process mainly relies on physical and chemical changes, covering a variety of technologies and methods. The most common desalination techniques include reverse osmosis, distillation, electrolysis, and membrane filtration. Reverse osmosis is one of the most widely used desalination technologies. It uses a semipermeable membrane to apply pressure to seawater, allowing water molecules to pass through while blocking larger salt ions and impurities. The efficiency of reverse osmosis depends on factors like membrane pore size, material, and applied pressure. Due to its relatively low energy consumption and high freshwater yield, reverse osmosis dominates many desalination plants. Another common technique is distillation, where seawater is heated to create steam, which is then condensed back into freshwater. Traditional distillation methods are effective but energy intensive. Modern versions, like multi-effect distillation (MED) and multi-stage flash distillation (MSF), improve efficiency by evaporating and condensing water in multiple stages, significantly increasing the output of fresh water while reducing energy consumption. Electrolysis uses electrical currents to cause a chemical reaction in seawater, separating it into hydrogen, oxygen, and leaving behind freshwater. Although it has been demonstrated in laboratory settings, its high energy demands and equipment costs prevent widespread commercial use. Membrane filtration removes most salt and impurities by physical filtration, employing techniques such as microfiltration, ultrafiltration, and nanofiltration to precisely separate components in seawater. However, membrane technology faces challenges like membrane fouling and limited lifespan. In addition to these technologies, there is growing interest in renewable energy-powered methods such as solar distillation and air evaporation. These methods harness natural solar energy or air humidity to extract fresh water through evaporation and condensation processes, offering environmentally friendly and sustainable alternatives. In summary, seawater desalination works by utilizing various physical and chemical processes to continuously remove salt and impurities, ultimately converting seawater into freshwater [6].

2.2 Solar desalination system

Precipitation is the primary source of freshwater globally, and the natural hydrological cycle that leads to precipitation can be seen as the foundational model for solar seawater desalination. Both academic research and industry practice have shown that thermal desalination processes are more suitable for large-scale applications compared to mechanical desalination processes. Solar energy is particularly well-suited for driving thermal desalination processes because it can be directly applied without any energy conversion losses. Moreover, existing thermal desalination installations can be easily retrofitted to accommodate solar energy. This makes solar energy an excellent choice for powering these processes, as shown in **Figure 2** [7].

Solar energy can be directly converted or otherwise utilized for desalination. There are two types of collection systems: direct desalination and indirect desalination in **Figure 2**. Direct desalination systems use solar energy to produce distilled water right in the container of the solar collector itself. Furthermore, indirect systems combine solar collection with traditional seawater desalination methods. Solar energy

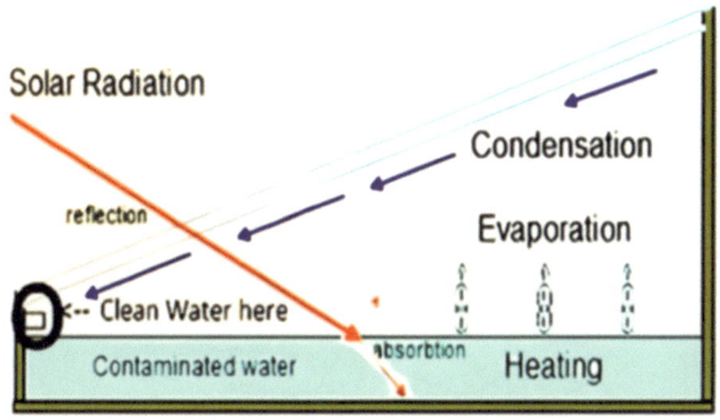

Figure 2.
Working principle of solar desalination system [7].

generates heat for desalination, providing power to desalination plants. In resource-scarce areas, like deserts or places with brackish water, renewable energy obtained from oil and gas is used for desalinating seawater [8].

2.2.1 Direct desalination system

2.2.1.1 Solar distillation

Solar distillers are eco-friendly, using solar power to desalinate brine instead of fossil fuels. However, the heat and flux from the sun can be a bit unpredictable. There are mainly two types of solar distillers: passive and active. The passive solar distiller is a cost-effective way to get drinking water or distilled water. With heat transfer, the raw water evaporates, leaving behind sediment and salt deposits in the basin; when the vapor moves to the relatively cooler glass cover, it releases latent heat and turns back into pure liquid through condensation. One downside of passive solar distillation is that it does not produce enough fresh water per square meter. An active solar distiller has a glass cover, a basin, a reflector, and a pump. The glass cover, flat reflector, and basin insulation have different thicknesses. Sunlight keeps hitting the reflector, heating up the brackish water. This preheated water then goes into the distiller. As the sun's rays hit, the water evaporates and collects on the glass cover, as shown in **Figure 3** [9]. In this system, the water to be distilled is introduced into the distillation device through a pipeline from the feed water tank. When sunlight penetrates the inclined glass cover, the energy it carries is absorbed by the water in the distillation device, causing the temperature of the water to gradually increase. Water molecules gain enough energy to overcome the intermolecular forces of the liquid state and then transform into a gaseous state to form water vapor. During the upward movement of these water vapors, they encounter the relatively low-temperature inner surface of the glass cover. Due to the temperature difference between the glass cover and the external environment, condensation will occur on its surface, and the water vapor will be transformed back into liquid water to form condensed water droplets. Under the action of gravity, these water droplets slide down along the glass cover onto the internally inclined panel and finally flow into the distilled water collection container along the panel. Thus, the separation process from raw water to distilled water is completed. Impurities in the raw water are left at the bottom of the distillation device, realizing the purification and distillation of water. Despite the numerous merits of the solar

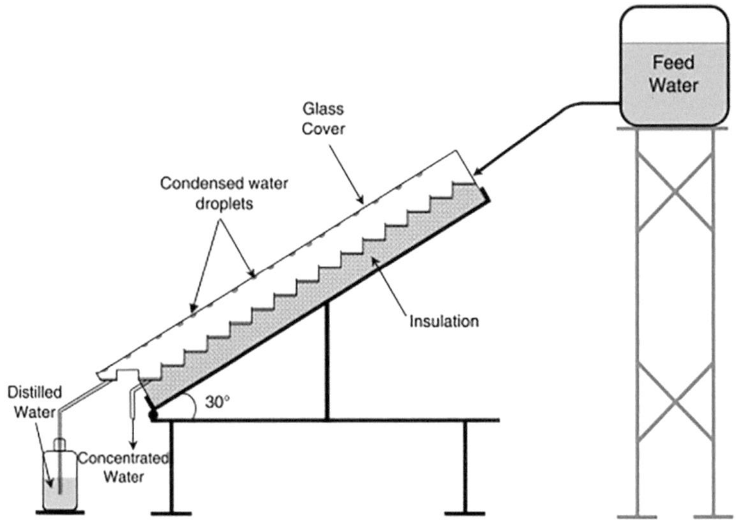

Figure 3.
Simple solar direct distillation.

direct distillation method, it is still confronted with a series of formidable challenges that urgently demand solutions in practical applications. On the one hand, the issue regarding its efficiency is rather conspicuous. Both the energy conversion efficiency and the evaporation-condensation efficiency leave much to be desired. A substantial amount of solar radiation energy is dissipated due to factors like reflection and heat dissipation. Furthermore, there exist efficiency bottlenecks in the steam condensation process, which consequently leads to a relatively low proportion of effective energy actually harnessed for water evaporation. On the other hand, this method entails a relatively large footprint. The meager water production hinges on extensive spatial requirements, and the system layout is rather dispersed. In the current context where land resources are constrained, this significantly curtails its large-scale dissemination. Simultaneously, climatic conditions also exert a pronounced impact on it. During overcast days, at night, or under extreme weather circumstances, the instability or absence of solar radiation will trigger a sharp decline in water production and may even inflict damage on the system, thereby severely undermining the stability and continuity of its operation. In addition, the vulnerability of water quality to contamination cannot be overlooked. Whether it be the leachates from the materials of the distiller or the microorganisms proliferating within, they will all give rise to the deterioration of water quality, rendering it difficult to meet potable water standards. Nevertheless, there are viable strategies to tackle these challenges. To enhance the efficiency, we can commence with optimizing the materials and structural design. Incorporating novel light-absorbing materials such as nano-coatings and refining the steam channels, while simultaneously integrating auxiliary energy sources or energy storage devices, can ensure the continuous and efficient progress of the distillation process [10]. With respect to reducing the occupied area, we should vigorously explore and develop high-efficiency and compact multi-layer or folding distillers and embrace the design concepts of integration and modularization to augment the freshwater yield per unit area. To cope with the limitations imposed by climatic conditions, we can combine energy storage technologies with backup energy systems and fortify the system's resilience against disasters to guarantee its stable operation under diverse weather conditions. For the matter of safeguarding water quality, it is imperative to select appropriate nonpolluting materials and reinforce the procedures of water quality monitoring and disinfection treatment, thereby ensuring the quality and safety of the produced freshwater. Through the comprehensive implementation of these countermeasures, it is anticipated that the current predicaments faced by the solar direct distillation method can be surmounted, facilitating its extensive application and long-term development in the domain of water resources.

2.2.1.2 Humidification-dehumidification process

The Humidification-Dehumidification (HD) process is based on the principle that air can mix with a large amount of water vapor. In this process, flowing air passes close to saline water, where it absorbs a certain amount of vapor from the air, inducing cooling. Conversely, distilled water can be recovered by bringing humid air into contact with a cold surface, causing condensation. HD technology is more suitable for desalination applications where water demand is not concentrated. The advantages of this method include reliable capacity, ease of use, and economical operating costs [11]. The principle of the humidification and dehumidification process is well reflected in **Figure 4**. The system first initiates the heating process. The inlet hot water is introduced into the heater and undergoes efficient heat exchange with the heating steam. The heat of the steam is accurately transferred to the inlet hot water, causing its temperature to increase significantly. The hot water that has completed the heating task is then transformed into outlet cold water and flows out, while the thermal energy of the heating steam is fully utilized to preheat for the subsequent seawater desalination process. Next comes the humidification stage. The preheated inlet seawater flows into the humidifier, and at the same time, the outside air is drawn in. Inside the humidifier, the special packing materials greatly increase the contact area and contact time between the seawater and the air, enabling the seawater to evaporate fully. The water vapor is incorporated into the air, resulting in a substantial

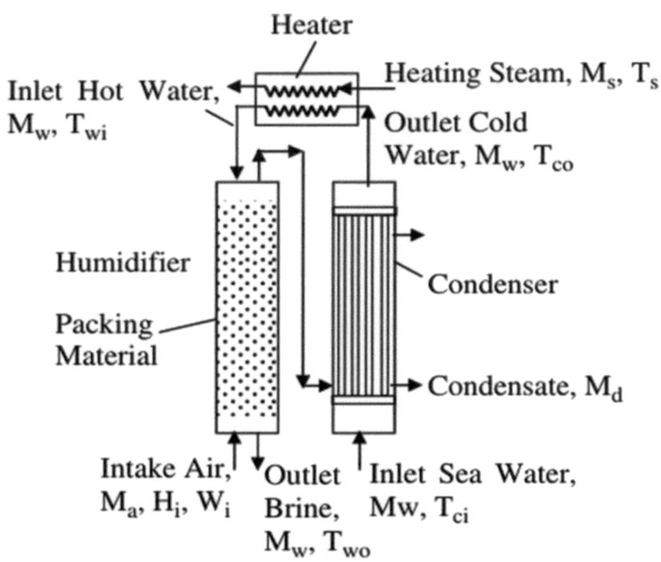

Figure 4.
Conventional humidification-dehumidification desalination process [12].

increase in humidity, while the seawater is concentrated into brine and discharged. The high-humidity air rich in water vapor then enters the condenser. Through the effect of the temperature difference with the external environment, the condenser rapidly cools the air. The water vapor in it condenses into liquid condensate on the inner surface of the condenser when encountering the cold. These condensates are efficiently collected, and the dehumidified air is finally discharged from the system. Simultaneously, the intermittent and unstable nature of solar energy poses a challenge in maintaining a consistent and steady power supply for seawater desalination systems. To address this, state-of-the-art energy storage devices can be deployed, such as intelligent battery management systems and highly efficient thermal storage facilities, which are capable of storing surplus energy to ensure uninterrupted operation. The issues of scaling and corrosion on equipment significantly impact their service life and operational efficiency. Opting for novel anti-corrosive and anti-scaling materials, in conjunction with regular physical and chemical cleaning and maintenance procedures, can effectively mitigate these problems. However, the high equipment costs have limited its wide application. We often choose to optimize production technologies and develop low-cost yet high-performance components, such as solar collectors with enhanced absorption capabilities and advanced humidification and dehumidification devices [13].

2.2.1.3 Solar chimneys

The materials for solar chimneys are usually readily available, and they require little maintenance, as well as no fuel or cooling water. However, solar chimneys are vulnerable to environmental issues such as dust or abrasion. In **Figure 5**, a key disadvantage of solar chimneys is the low efficiency of converting solar thermal energy into electricity [14]. Consequently, it is advisable to implement a high-efficiency air filtration system at the ingress of the chimney. Specifically, a combination of

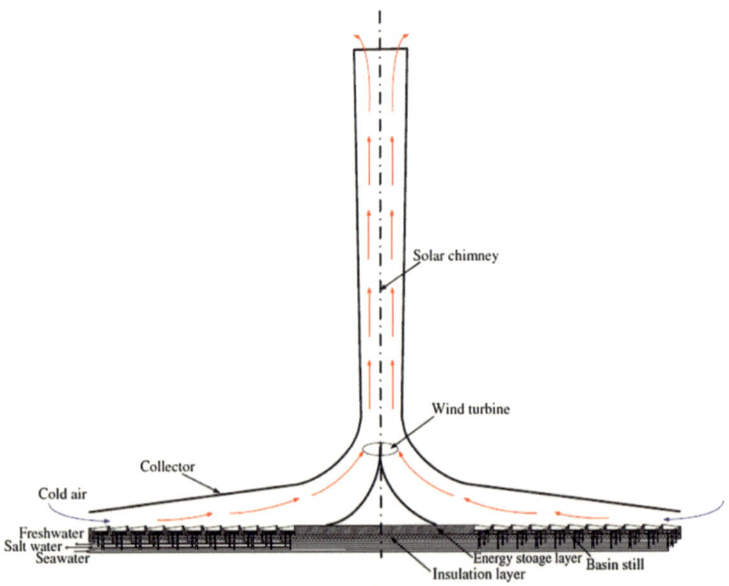

Figure 5.
Principle of solar chimney [14].

multi-layered electrostatic precipitator filters and coarse filters can be employed, which is capable of effectively impeding the ingress of dust particles and mitigating the abrasion and erosion on the inner surface of the chimney. Simultaneously, novel ceramic coating materials with excellent wear and corrosion resistance properties can be utilized on the inner wall. These materials not only augment the wear resistance but also confer protection against the corrosive effects of salts present in the seawater evaporation environment, thereby extending the operational lifespan of the chimney.

2.2.2 Indirect desalination systems

2.2.2.1 Solar photovoltaic-reverse osmosis

Countries with a per capita annual water consumption of 1000–1700 cubic meters are considered water-scarce. Excessive consumption by each citizen or individual will exacerbate the water crisis, worsening water-related problems. Reverse osmosis seawater desalination plants use seawater as their source and increase water pressure through electric-powered pumps. The water travels through pipes with semipermeable membranes inside. These membranes block the passage of salt, allowing the water to move forward. The water is split into two parts in **Figure 6**: one with high salt concentration and the other with low salt concentration, referred to as the permeate or product water. In reverse osmosis, the membrane is the most crucial element and must have the following characteristics: (1) It must withstand operational conditions; (2) It must reject a high percentage of salt to provide high-quality output; (3) It should have sufficient permeability to maintain a high flow of supplied water [16]. Meanwhile, we can organically combine solar energy with reverse osmosis, such as the solar photovoltaic-reverse osmosis technology [17]. It integrates the advantages of solar power generation and reverse osmosis membrane separation, demonstrating a unique working principle. Solar photovoltaic panels convert solar light energy into electrical energy through the photoelectric effect, providing power support for the high-pressure pumps, control systems and so on in the reverse osmosis system. During the reverse osmosis process, seawater undergoes pretreatment to remove impurities first and then is pressurized by the high-pressure pump to a pressure higher than its osmotic pressure, forcing water molecules to pass through

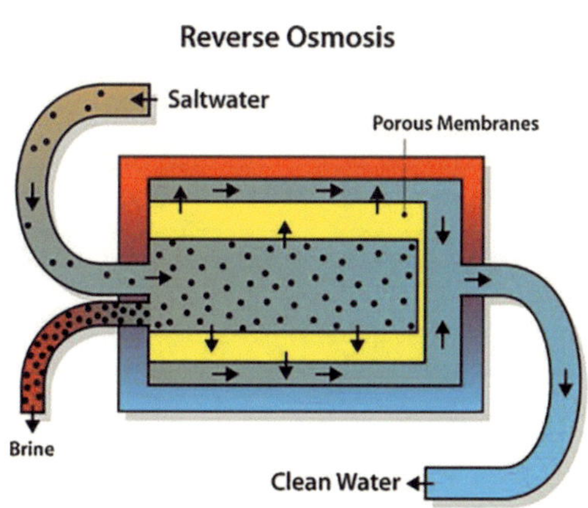

Figure 6.
Reverse osmosis desalination technology [15].

the semipermeable membrane to form fresh water, while impurities such as salts are retained and discharged. However, this technology faces numerous challenges. The intermittence and instability of solar energy lead to fluctuations in power supply, which affects the continuous operation of the system. To address this, high-efficiency energy storage devices can be equipped, such as intelligent battery management systems and new heat storage equipment, to store excess energy and ensure a stable energy supply. The high cost of equipment limits its wide application. The manufacturing costs of solar photovoltaic panels and reverse osmosis equipment should be reduced through technological innovation and large-scale production. Meanwhile, the system design should be optimized to improve the cost-performance ratio. In addition, the salt scale and microorganisms in seawater are likely to cause blockage and pollution of the reverse osmosis membrane, reducing the membrane's lifespan and desalination efficiency. Therefore, new anti-fouling and anti-scaling membrane materials can be adopted.

2.2.2.2 Multi-effect distillation

The multi-effect distillation seawater desalination technology is based on the principle of multiple utilization of steam, ingeniously achieving efficient desalination of seawater. In the process shown in **Figure 7**, seawater first enters the preheater and is preheated by utilizing the latent heat of condensation of the steam generated in each effect. Then it enters the first-effect evaporator. An external heat source (such as steam or waste heat) makes the seawater boil and evaporate in the first-effect evaporator. The secondary steam generated enters the second-effect evaporator as a heating heat source, and so on. The subsequent effect evaporators use the secondary steam generated by the previous effect to carry out evaporation operations one by one. The pressure and boiling point in each effect evaporator gradually decrease, thus realizing the multiple utilization of steam and greatly improving the energy utilization efficiency. Finally, fresh water is obtained through condensation. However, there are also some problems at the same time. The problem of equipment scaling is rather prominent. Calcium, magnesium and other ions in seawater are prone to precipitate and form scale layers during the evaporation process, which adhere to the heat exchange surfaces of the evaporators, reducing the heat transfer efficiency, increasing energy consumption and even possibly causing pipeline blockages. To overcome this

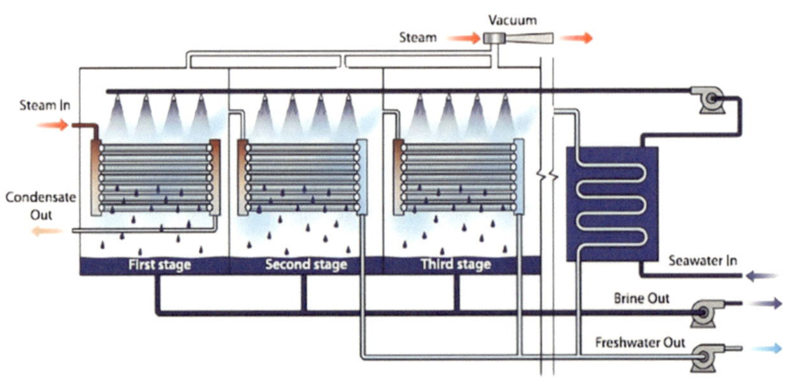

Figure 7.
Multi-effect distillation technology [18].

problem, pretreatment techniques can be adopted, such as adding scale inhibitors and conducting ion exchange to soften the seawater. Meanwhile, regular chemical cleaning and physical flushing are carried out to ensure the cleanliness of the heat exchange surfaces. Then, the initial investment cost of the multi-effect distillation system is relatively high, including the purchase and installation costs of equipment such as evaporators, condensers, vacuum pumps, as well as the costs of factory building and so on, which restricts its promotion and application to a certain extent. By optimizing the system design and adopting modular and standardized production methods to improve the integration and manufacturing process level of the equipment, it is expected to reduce the equipment cost [19].

2.2.2.3 Multi-stage flash evaporation

The desalination technology continuously evaporates seawater in multiple flash chambers under different pressures using the thermal energy provided by solar power in **Figure 8**. In this process, preheated seawater rapidly evaporates (flashes) in a low-pressure environment. The resulting steam condenses into freshwater in a condenser, and the recovered heat is used for subsequent evaporation. Multi-stage flash technology works through cycles of evaporation and condensation. Solar-powered multi-stage flash plants can produce 6–60 liters of water per square meter per day. Heated seawater is introduced into a low-pressure space, where the ambient pressure is lower than the saturated vapor pressure corresponding to the temperature of the heated seawater. As a result, the seawater rapidly partially evaporates to produce steam, which is then condensed into freshwater [20]. At present, this particular technology is beset by the disadvantage of a relatively elevated energy consumption level. The issues of equipment scaling and corrosion have significantly eroded stability and curtailed the service life of the overall system. The initial investment outlay for the multi-stage flash system is exorbitantly high, encompassing a wide spectrum of expenditures from the acquisition of large-scale and intricate equipment, and the establishment of manufacturing facilities, to the painstaking installation of pipelines. In order to surmount these impediments, the adoption of a new generation of high-efficiency heat exchangers is warranted. This would enhance the heat recovery efficiency and mitigate heat dissipation. Concurrently, the pretreatment procedures for seawater

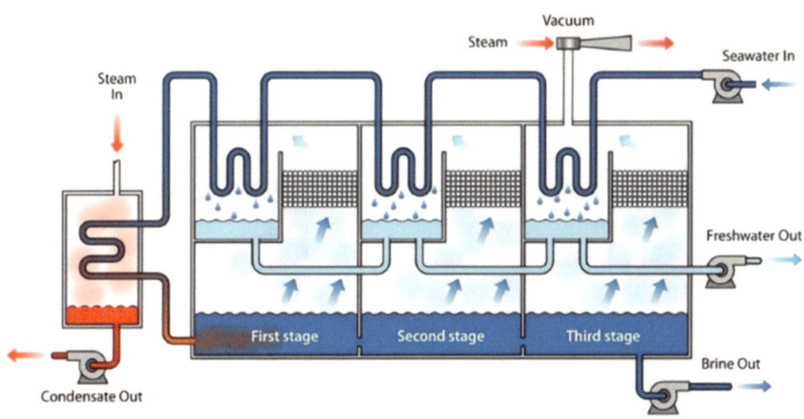

Figure 8.
Multi-stage desalination technology [18].

ought to be intensified. By leveraging advanced methodologies such as filtration, ion exchange, and the addition of chemical reagents, the removal of deleterious impurities can be optimized to a greater extent.

2.2.2.4 Analysis of different solar desalination technologies

Approximately 61% of the world's population has access to drinking water. However, in rural areas, drinking water is likely to be contaminated. Using large reservoirs to purify water often comes at the cost of harming nature. In terms of solar-powered methods, eco-friendly water purification is cost-effective, attractive, and essentially trouble-free. Solar desalination is a unique process that uses freely available solar energy as a heating medium. Desalination of seawater can only become a more popular freshwater supply option without significantly reducing energy and costs. Therefore, in places where water resources are scarce and water costs are high, considering the feasibility of desalination is most appropriate. Moreover, higher water temperatures and significantly lower salinity can greatly reduce the energy intensity of membrane desalination. However, desalination is not a magic solution to water supply problems; it does not address poor sanitation, weak management systems, or insufficient distribution.

Dai and Zhang [21] conducted a solar desalination experiment related to the humidification-dehumidification process at Northwestern Polytechnical University. They found that the system's output is highly correlated with the temperature of the brine entering the humidifier, the mass flow rate of the brine, and the mass flow rate of process air. The thermal efficiency of the device exceeded 80%. As a result, waste heat can be reused to further enhance the desalination process. This system can be used for seawater desalination.

Kabeel and El-Said [22] studied a hybrid solar desalination device, which includes a humidification-dehumidification (HD) unit and a single-stage flash device. Their main goal was to investigate this system. The results showed that the device has significant operational flexibility between air HD and flash desalination. Experimental results indicated a considerable increase in the production efficiency of the HDH unit.

Joseph et al. [23] conducted a study on a single-stage solar desalination system. The goal of this study was to produce 10 liters of drinking water per day. The results showed that the water quality was acceptable and within tolerance limits. Additionally, the study indicated that the device produced approximately 8.5 liters of water per day, which is three times the output of a typical solar distiller.

In summary, desalination can be roughly divided into direct methods and indirect methods, as shown in **Figure 9**. Direct methods include solar distillation, humidification and dehumidification, and solar chimney technologies, which utilize solar radiation to desalinate seawater directly. Solar distillation is an ancient and effective method that involves heating seawater using solar energy, causing it to evaporate and then condense into fresh water. The humidification and dehumidification method extracts fresh water by controlling the humidity of air. The solar chimney creates an updraft to promote water evaporation and condensation into fresh water. Furthermore, indirect methods convert solar energy into other forms of energy to desalinate seawater, including reverse osmosis, multi-effect distillation, and multi-stage flash distillation technologies. Reverse osmosis employs solar energy to drive pump systems, pushing seawater through semipermeable membranes to remove salt

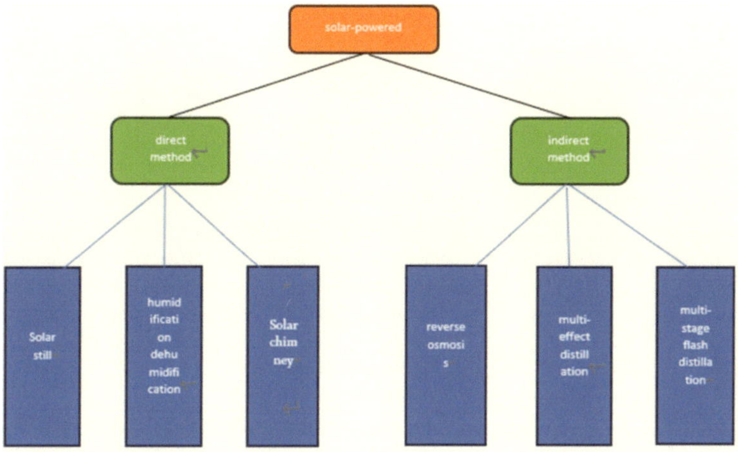

Figure 9.
Solar desalination classification.

and impurities. Multi-effect distillation and multi-stage flash distillation utilize the heat generated by solar energy to extract fresh water through multiple heating and condensation processes.

3. Analysis of the benefits of solar desalination

The main economic advantages of a solar-powered seawater desalination system are as follows: (1) It does not require much infrastructure. (2) It is easy to design, install, operate, and maintain locally. Furthermore, its main economic disadvantage is the limitation on unit size due to the large area required (about 250 m^2 per 1 m^3 of freshwater). Therefore, based on different solar desalination methods, the goal is to study the cost of producing 1 liter or 1 cubic meter of freshwater, as well as to analyze the economic benefits by calculating the proportion of each cost in the total cost. However, the feasibility and cost-effectiveness of desalination plants depend on the specific location and various other factors (**Table 1**) [31].

Hamed [32] believes that the cost of seawater desalination varies depending on the desalination technology used and the water supply demand. Since most thermal desalination plants use coal and oil, the unit cost of desalinated water is estimated to be quite high. The cost of desalinating seawater ranges from $0.521 to $1.02 per cubic meter, while multi-effect distillation (MED) plants have a high capacity of over 91,000 m^3/day. For MED plants with a daily output of 12,000 m^3 to 50,000 m^3, the water cost ranges from $0.960 to $1.5/$m^3$, whereas for MED plants with a capacity of less than 100 m^3/day, the unit cost per cubic meter of water ranges from $2 to $10/$m^3$ (**Table 2**).

According to the research by Karagiannis et al. [34], the Total Dissolved Solids (TDS) in brackish and freshwater rise between 2000.0 and 10,000 ppm, affecting the cost of water supply. For instance, brackish water in Jordan containing 2300.0 ppm has a cost of $0.26 per cubic meter, while water with 5000 ppm has a cost of $0.27 per cubic meter. Therefore, in two equivalent systems, the volume of water depends on the total TDS. When using sustainable sources, the cost of freshwater treatment

	Type of technology	Type of water supply	Specificities	Cutting edge	Quote
1	New Heat Absorber Triangle Solar Distiller	Sea water	By using a new form of trapezoidal absorber to reduce the effect of less shadowing produced by the sloping side walls and volume of seawater	Up to half the efficiency compared to plate-constructed stills	[24]
2	New integrated multi-effect distillation-absorber compressor (MED-AB)	Sea water	In addition to the additional steam generated in the absorption tower, the final steam is recovered through the AB system	MED-AB's operating costs are about 38% lower than conventional multi-effect distillation (MED)	[25]
3	A new renewable energy source combined with solar energy-driven reverse osmosis desalination with zero liquid discharge	Seawater/ saltwater	The saline waste from the RO (Reverse Osmosis) process is remixed with water from the evaporator to reduce the salinity of the RO feed. The hydrogen produced in the electrolytic cell is burned, and its energy is stored in cement blocks. The evaporator operates using solar energy during sunny hours of the day. Then, the energy stored in the cement blocks replaces solar energy during the remaining hours of the day.	Greatly reduce the cost of seawater desalination	[26]
4	New nanofluids for solar distillation.	Seawater/ saltwater	Using nanofluids for volumetric heating in overall seawater heating can generate localized high temperatures around the metal nanoparticles within the fluid, increasing the heating rate and enhancing the evaporation process	The dispersed metal or carbon nanoparticles in the liquid absorb sunlight and generate steam without needing to heat the entire fluid volume	[27]
5	A solar chimney power plant with solar desalination capabilities.	Seawater	The chimney acts as a negative pressure pipe, increasing the pressure differential.	Maximize the use of solar energy	[28]
6	The recycling of waste heat promotes solar desalination.	Seawater	Integration of solar-heated HDH cycle with diesel engine exhaust gases and discarded coolant flow	To optimize the freshwater production rate, recovery rate, specific entropy generation, and gained output ratio of a seawater desalination system while minimizing the waste heat discharged into the environment.	[29]

	Type of technology	Type of water supply	Specificities	Cutting edge	Quote
7	An integrated system based on solar chimneys and wind energy for hybrid desalination through reverse osmosis and multi-stage flash distillation with brine recovery.	Seawater	Using cascade desalination with thermal desalination and membrane desalination technologies, the heat stored in a solar chimney is utilized to produce fresh water. Additionally, a pressure-retarded osmosis subsystem is integrated to enhance desalination	Compared to traditional multi-stage flash distillation, it achieves more efficient energy use and higher freshwater production	[30]

Table 1.
Analysis chart of different solar desalination technologies.

Desalination methods	Capital cost range	Cost composition factors	Cost range and characteristics of per liter of water.
Indirect solar still	The capital cost ranges from 2.52 to 282 dollars. There is a large difference among different types (for example, the single-basin solar still has a lower cost while the sun-tracking solar still has a higher cost).	Mainly equipment and material costs (such as glass, metal plates, etc.), and a small amount of maintenance costs (regular cleaning and water replenishment).	Passive CPL (Cost Per Liter) does not exceed 0.1667 US dollars. The lowest is 0.0074 US dollars for a single-effect tubular solar still.
Direct solar still	The capital cost is between 160 and 1174 dollars. The higher capital cost mainly stems from additional energy equipment (such as flat plate collectors, vacuum tube collectors).	Equipment and collector costs, maintenance costs (involving the maintenance of additional equipment), and energy-related costs.	Active CPL does not exceed 0.2696 US dollars. The active type has a minimum of 0.0066 US dollars for the point-focusing elliptical solar still.

Table 2.
Comparison of cost situations of different seawater desalination technologies [33].

Typology	Renewable energy	Cost of production ($/m^3)
Seawater	Conventional fuel	0.46–3.5
	Tidal	1.3–6.5
	Photovoltaic cell	4.08–11.7
	Solar collector	4.5–10.4
Saltwater	Conventional fuel	0.27–1.38
	Photovoltaic cell	5.8–13.42
		2.61

Table 3.
Water production cost (US$/m^3) for Ocean water and brackish water by utilizing a variety of sources of energy [35].

is significantly higher. In some cases, it can exceed $10.3 per cubic meter, which is even more expensive than some of the priciest energy supply options. However, when desalinating seawater (SW) or brackish water (BW) using conventional energy sources, the costs are much lower (**Table 3**).

4. Commercialization development of solar thermal desalination

4.1 Direct solar desalination

Direct solar desalination technology has the advantages of relatively simple systems and low operation and maintenance costs and has good commercial application potential in some small and decentralized freshwater demand scenarios. For example, in some remote islands, coastal small villages, and other areas, small direct solar distillation devices can meet the basic domestic water needs of local residents. Their low costs and simple operation methods make these areas more likely to accept and adopt them. However, there are also some limitations. For instance, the energy conversion efficiency is relatively low. In large-scale desalination projects, a larger heat collection area is required to meet the water production requirements, which increases the floor space occupied by the equipment and the initial investment costs. Moreover, water production efficiency is greatly affected by weather conditions. In cases of insufficient sunlight such as cloudy and rainy days, the water production volume will drop significantly, resulting in insufficient stability of the water supply. Currently, there are already some practical application cases of direct solar desalination technology worldwide, but the overall scale is relatively small. In the coastal areas and islands of some developing countries, small solar distillation devices are providing domestic water for local residents [36].

4.2 Indirect solar desalination

Indirect solar desalination technology has been applied in some specific regions and scenarios. For example, in some countries in the Middle East, due to their abundant solar energy resources and the urgent need for fresh water, indirect solar desalination technology has been used to build medium-sized desalination plants.

Some of the technologies have achieved commercial production. For instance, membrane distillation technology, which is a common type of indirect solar desalination technology, has seen the development of some commercial modules. The plate-and-frame module constructed by Scarab AB, which consists of 10 cassette structures, has a certain commercial application value [37]. However, traditional reverse osmosis desalination technology and distillation desalination technology are more mature in large-scale commercial applications and occupy a relatively large share of the global desalination market. Indirect solar desalination technology has problems such as high equipment costs, relatively low energy conversion efficiency, and complex system maintenance, which lead to higher costs for desalinated water. As a result, it is at a disadvantage in commercial competition and restricts the speed of its large-scale commercial transformation.

4.3 Specific solar desalination technologies

The commercial applications of solar stills are mostly focused on small-scale projects. For example, in some remote islands, small communities in coastal arid areas, and special places like ships, direct solar stills are used to meet the basic freshwater needs of local residents or specific users. The daily water production varies from several hundred liters to thousands of liters [38]. For instance, some islands in Greece once adopted direct solar stills to provide domestic water for local residents. Its initial investment cost is relatively low. Especially for small and simple distillation devices, the costs of equipment procurement and installation are not high. For example, The equipment costs of some small solar tray stills may range from several hundred to several thousand dollars [39].

Solar photovoltaic-reverse osmosis, relatively low energy consumption, and high fresh water production make it a widely used desalination technology. However, the equipment cost is high. When solar energy is combined with reverse osmosis, the intermittency and instability of solar energy will affect the continuous operation of the system. Commercial feasibility can be improved by equipping high-efficiency energy storage equipment, reducing the manufacturing cost of solar photovoltaic panels and reverse osmosis equipment, optimizing system design, and adopting new anti-fouling and anti-scaling membrane materials. It is suitable for fresh water demands of different scales, especially for medium to large-scale applications.

The multi-effect distillation (MED) desalination technology has been applied to a certain extent worldwide, especially in regions with scarce water resources and abundant energy, such as the Middle East and North Africa, including countries like Saudi Arabia and the United Arab Emirates. However, in the global desalination market, the market share of MED technology is relatively limited. The high equipment investment and operating costs of MED technology have restricted its promotion and application in some areas. According to relevant data, in the global desalination market, the market share of multi-effect distillation (MED) technology is rather small, while the market share of the reverse osmosis (RO) technology exceeds 50% [40].

Multi-stage flash (MSF) technology is prevalently employed in large-scale seawater desalination endeavors and has witnessed extensive utilization in the Middle East region [41]. In locales such as Abu Dhabi, United Arab Emirates, expansive multi-stage flash seawater desalination and cogeneration systems have been erected, with the capacity of a single unit reaching up to 57,700 cubic meters per diem. Likewise, in Saudi Arabia and other nations, there exist numerous multi-stage flash seawater desalination plants capable of yielding hundreds of thousands of tons of fresh water daily [42], furnishing copious amounts of fresh water for urban water supply and industrial water demands. Nevertheless, the technology is encumbered by exorbitant construction costs. This is attributable to the requisite high-pressure apparatuses and corrosion-resistant materials, which incur substantial expenditures in equipment fabrication, procurement, and material acquisition. The operation and maintenance costs are also prohibitive, encompassing energy consumption, equipment upkeep, and chemical agent expenses. The multi-stage flash process is energy intensive and frequently necessitates integration with thermal power plants. Equipment maintenance demands the engagement of specialized personnel and significant financial outlays, thereby circumscribing the ambit of its commercial dissemination.

5. Future expectations

5.1 Hybrid rare earth-driven desalination

The application of solar energy to desalination processes shows potential for further growth in terms of improving performance, saving energy, or reducing costs.

In particular, the use of hybrid renewable energy systems to promote desalination is of great interest to the research community [43].

5.1.1 Reverse oasis with photovoltaic and wind energy

The advantage of a hybrid PV-wind system is that each source can compensate for the lack of availability of the other. This is relevant considering the variable nature of both and ensures more continuous energy availability. So far, there have been some studies that have optimized hybrid solar PV systems-wind power systems, taking into account variables such as water availability, load, location of solar wind farms, and their size [44]. In the only experimental study conducted in 2001 that combined photovoltaic-wind energy with desalination, Weiner et al. experimentally designed and operated a small, stand-alone brackish water reverse osmosis (BWRO) plant that combined solar photovoltaic and wind energy. Mokheimer et al. [45] simulated a small-scale hybrid wind/solar PV RO desalination system with a production capacity of 5 m^3/day. They found that the cost of a hybrid powertrain could be reduced if multiple wind turbines were used. While the efficiency of the entire system depends on feed salinity, this also determines the load demand. Similarly, Alberman and ElAmin investigated solar PV/wind/diesel hybrids in remote areas of Saudi Arabia. They found that despite the progress made in reducing PV costs over the past few decades, only diesel systems are the most cost-efficient. This shows that there is still untapped potential for renewables-driven desalination.

5.2 Other desalination processes

5.2.1 Positive infiltration

Forward osmosis is a salinity gradient-driven or gravity-driven desalination process in which the differential pressure across the semipermeable membrane results in selective water transport. Due to its ability to naturally diffuse water from the feed into a higher concentration solution, it does not require hydraulics, making it a low energy alternative to RO desalination. As a result, it has the potential to significantly reduce the energy costs associated with desalination. Recently there have been several FO units, most of them in China, and one of them has a capacity of 2800 m^3/day, the highest capacity of any FO unit installed to date. Khaydarov [46] introduced the concept of solar FO in 2007, using various fluids of a solar cell system, a solar heat exchanger, a pretreatment unit, and FO equipment. Razmjou et al. [47] investigated the feasibility of bilayer polymer hydrogels as extractants for FOs using solar energy concentration. The hydrogel consists of an absorbent layer that provides osmotic pressure and a dehydrated layer that absorbs water during the release of FO in **Figure 10**.

5.2.2 Dew point evaporation

The future trends in solar seawater desalination include a shift toward less-explored desalination technologies. Dew point evaporation is one such technique with the potential for solar desalination. In dew point evaporation, saturated vapor is used as a carrier gas to evaporate water from the saline feed as distillate. This process offers the advantage of energy reuse, depending on the use of heat exchangers and improved process design [48].

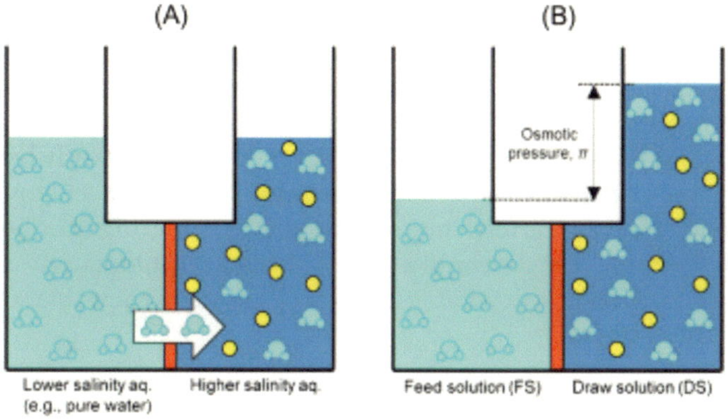

Figure 10.
Schematic of FO process [47].

6. Conclusions

Solar-powered desalination is a rapidly growing research area that has made significant strides in recent years. The increasing desalination capacity, combined with the competing demands for decarbonization and mitigating the adverse effects of global warming, has driven efforts to use renewable energy for desalination. Solar energy, in particular, is an attractive option because of its potential to power desalination, especially in regions where freshwater scarcity and solar radiation coexist.

The interest in direct solar desalination has surged due to advancements in novel photothermal materials. Graphene-based and metal/ceramic nanostructures have facilitated evaporation through localized heating. Current leading solar desalination technologies include solar distillation, reverse osmosis, multi-stage flash distillation, multi-effect distillation, and solar chimneys. Each of these technologies has its own advantages and disadvantages, tailored to different scales and demand scenarios.

Solar distillation is suitable for small-scale, distributed systems and is especially useful in remote, resource-scarce regions. On the other hand, reverse osmosis, with its higher energy efficiency and water production capacity, is more appropriate for medium to large-scale applications. Multi-stage flash and multi-effect distillation technologies, known for their efficient energy usage, are typically employed in large industrial facilities.

From a cost-effectiveness perspective, while the initial investment for solar desalination technologies is relatively high, ongoing technological advancements and the promotion of large-scale applications have gradually reduced the cost per unit of water produced, with operational costs becoming more manageable. In regions rich in solar resources, these technologies demonstrate strong economic feasibility. Additionally, as global attention on low-carbon economies and renewable energy grows, solar desalination technologies are expected to receive more policy support, further lowering economic barriers.

In terms of environmental and social benefits, solar desalination technologies can significantly reduce greenhouse gas emissions compared to traditional fossil fuel-powered desalination methods, thus minimizing environmental impact. Moreover, these technologies can effectively alleviate water scarcity in many regions, particularly in arid and semi-arid areas, where their social benefits are especially pronounced. By providing a stable supply of freshwater, solar desalination technologies can improve the life quality of residents and support regional economic development.

Future developments will focus on improving efficiency, reducing costs, and achieving large-scale applications. Researchers are exploring the integration of advanced materials, smart control systems, and energy storage technologies with existing solar desalination systems to enhance efficiency and system reliability. For instance, the application of nanomaterials is expected to significantly boost photo-thermal conversion efficiency and evaporation rates. Meanwhile, energy storage technologies can mitigate the intermittency of solar energy, ensuring continuous system operation. Furthermore, the development of multi-energy integration systems, such as combining solar energy with wind, geothermal, or other renewable sources, is a key area for future research.

Conflict of interest

The authors declare no conflict of interest.

Author details

Huijin Xu* and Hao Luo
School of Merchant Marine, Shanghai Maritime University, Shanghai, China

*Address all correspondence to: hjxu@shmtu.edu.cn

References

[1] Mekonnen MM, Hoekstra AY. Four billion people facing severe water scarcity. Science Advances. 2016;**2**(2):e1500323

[2] Fang X, Wang J. Nonequilibrium thermodynamics in cell biology: Extending equilibrium formalism to cover living systems. Annual Review of Biophysics. 2020;**49**(1):227-246

[3] Kariman H, Shafieian A, Khiadani M. Small scale desalination technologies: A comprehensive review. Desalination. 2023;**567**:116985

[4] Esfahani IJ et al. Efficient thermal desalination technologies with renewable energy systems: A state-of-the-art review. Korean Journal of Chemical Engineering. 2016;**33**:351-387

[5] Arunkumar T et al. Productivity enhancements of compound parabolic concentrator tubular solar stills. Renewable Energy. 2016;**88**:391-400

[6] Ali MT, Fath HES, Armstrong PR. A comprehensive techno-economical review of indirect solar desalination. Renewable and Sustainable Energy Reviews. 2011;**15**(8):4187-4199

[7] Kalidasan B et al. Technological advancement of solar thermal system desalination process – A review. IOP Conference Series: Materials Science and Engineering. 2021;**1059**(1):012061

[8] Farid M et al. Solar desalination with a humidification-dehumidification cycle: Mathematical modeling of the unit. Desalination. 2003;**151**(2):153-164

[9] El-Agouz S et al. Performance evaluation of a continuous flow inclined solar still desalination system. Energy Conversion and Management. 2015;**101**:606-615

[10] Thakur AK et al. Synergetic effect of absorber and condenser nano-coating on evaporation and thermal performance of solar distillation unit for clean water production. Solar Energy Materials and Solar Cells. 2022;**240**:111698

[11] Abdelmoez W et al. Water desalination using humidification/dehumidification (HDH) technique powered by solar energy: A detailed review. Desalination and Water Treatment. 2014;**52**(25-27):4622-4640

[12] Ettouney H. Design and analysis of humidification dehumidification desalination process. Desalination. 2005;**183**(1-3):341-352

[13] Ghazouani N et al. Solar desalination by humidification–dehumidification: A review. Water. 2022;**14**(21):3424

[14] Maia CB et al. An overview of the use of solar chimneys for desalination. Solar Energy. 2019;**183**:83-95

[15] Al-Karaghouli A et al. Solar and wind opportunities for water desalination in the Arab regions. Renewable and Sustainable Energy Reviews. 2009;**13**(9):2397-2407

[16] Shalaby S et al. Reverse osmosis desalination systems powered by solar energy: Preheating techniques and brine disposal challenges–a detailed review. Energy Conversion and Management. 2022;**251**:114971

[17] He W, Wang Y, Shaheed MH. Stand-alone seawater RO (reverse osmosis) desalination powered by PV (photovoltaic) and PRO

(pressure retarded osmosis). Energy. 2015;**86**:423-435

[18] Al-Karaghouli A, Kazmerski LL. Energy consumption and water production cost of conventional and renewable-energy-powered desalination processes. Renewable and Sustainable Energy Reviews. 2013;**24**:343-356

[19] Liponi A et al. Multi-effect distillation plants for small-scale seawater desalination: Thermodynamic and economic improvement. Energy Conversion and Management. 2020;**205**:112337

[20] El-Dessouky HT, Ettouney HM, Al-Roumi Y. Multi-stage flash desalination: Present and future outlook. Chemical Engineering Journal. 1999;**73**(2):173-190

[21] Dai Y, Zhang H. Experimental investigation of a solar desalination unit with humidification and dehumidification. Desalination. 2000;**130**(2):169-175

[22] Kabeel A, El-Said EM. A hybrid solar desalination system of air humidification–dehumidification and water flashing evaporation: Part I. A numerical investigation. Desalination. 2013;**320**:56-72

[23] Joseph J, Saravanan R, Renganarayanan S. Studies on a single-stage solar desalination system for domestic applications. Desalination. 2005;**173**(1):77-82

[24] Boudhiaf R et al. Experimental analysis of triangular solar distiller with a new form of absorber. Desalination and Water Treatment. 2024;**317**:100087

[25] Aly S et al. Pilot testing of a novel integrated multi effect distillation - absorber compressor (MED-AB) technology for high performance seawater desalination. Desalination. 2022;**521**:115388

[26] Gadalla MA, Abdel Fatah A, Elazab HA. A novel renewable energy powered zero liquid discharge scheme for RO desalination applications. Case Studies in Chemical and Environmental Engineering. 2023;**8**:100407

[27] Neumann O et al. Solar vapor generation enabled by nanoparticles. ACS Nano. 2012;7(1):42-49

[28] Zuo L et al. Comprehensive study of wind supercharged solar chimney power plant combined with seawater desalination. Solar Energy. 2018;**166**:59-70

[29] Shamet O et al. Performance of different HDH desalination units powered by diesel engine generator waste heat. Process Safety and Environmental Protection. 2023;**179**:651-666

[30] Méndez C, Bicer Y. Integrated system based on solar chimney and wind energy for hybrid desalination via reverse osmosis and multi-stage flash with brine recovery. Sustainable Energy Technologies and Assessments. 2021;**44**

[31] Reddy KV, Ghaffour N. Overview of the cost of desalinated water and costing methodologies. Desalination. 2007;**205**(1-3):340-353

[32] Hamed OA. Overview of hybrid desalination systems—Current status and future prospects. Desalination. 2005;**186**(1-3):207-214

[33] El-Bialy E et al. Cost analysis for several solar desalination systems. Desalination. 2016;**384**:12-30

[34] Karagiannis IC, Soldatos PG. Water desalination cost literature: Review and assessment. Desalination. 2008;**223**(1-3):448-456

[35] Bourouni K, Chaibi MJGW. Application of geothermal energy for brackish water desalination in the south of Tunisia. Environmental Science, Engineering. 2005;**2185**(290):225

[36] Ayoub J, Alward R. Water requirements and remote arid areas: The need for small-scale desalination. Desalination. 1996;**107**(2):131-147

[37] Cipollina A et al. Development of a membrane distillation module for solar energy seawater desalination. Chemical Engineering Research and Design. 2012;**90**(12):2101-2121

[38] Galindo DM. Design of a Solar-Powered Desalination Plant for a Remote Kenyan Community. Barcelona City, Catalonia, Spain: Universitat Politècnica de Catalunya; 2024

[39] Eibling J, Talbert S, Löf G. Solar stills for community use—Digest of technology. Solar Energy. 1971;**13**(2):263-276

[40] Anis SF, Hashaikeh R, Hilal N. Reverse osmosis pretreatment technologies and future trends: A comprehensive review. Desalination. 2019;**452**:159-195

[41] Al-Addous M et al. Innovations in solar-powered desalination: A comprehensive review of sustainable solutions for water scarcity in the Middle East and North Africa (MENA) region. Water. 2024;**16**(13):1877

[42] Mir MA, Ashraf MW. Monitoring, and management, the challenges and potential strategies of Saudi Arabia's water resources: A review in analytical way. Environmental Nanotechnology Monitoring and Management. 2023;**20**:100855

[43] Khan MA et al. A hybrid renewable energy system as a potential energy source for water desalination using reverse osmosis: A review. Renewable and Sustainable Energy Reviews. 2018;**97**:456-477

[44] Cherif H, Belhadj JJE. Large-scale time evaluation for energy estimation of stand-alone hybrid photovoltaic–wind system feeding a reverse osmosis desalination unit. Fuel and Energy Abstracts. 2011;**36**(10):6058-6067

[45] Mokheimer EM et al. Modeling and optimization of hybrid wind–solar-powered reverse osmosis water desalination system in Saudi Arabia. Energy Conversion and Management. 2013;**75**:86-97

[46] Khaydarov RA, Khaydarov RR. Solar powered direct osmosis desalination. Desalination. 2007;**217**(1-3):225-232

[47] Razmjou A et al. Bifunctional polymer hydrogel layers as forward osmosis draw agents for continuous production of fresh water using solar energy. Environmental Science and Technology. 2013;**47**(22):13160-13166

[48] Pandelidis D et al. Water desalination through the dewpoint evaporative system. Energy Conversion and Management. 2021;**229**:113757

Chapter 4

Membrane Separation Processes: Principles, Structures, Materials, and Future Prospects

Mahmoud Shaban, Ashraf Morsy, Amira S. Mahmoud and Hamida Abdel-Hamid

Abstract

This chapter encompasses a thorough exploration of membrane separation processes, membrane structure, synthesis techniques, membrane materials, and membrane preparation techniques. It begins with an in-depth examination of the principles underlying membrane separation processes and their diverse applications across various industries, emphasizing their role in addressing environmental challenges and resource sustainability. Furthermore, it delves into the intricate details of membrane structure and synthesis techniques, elucidating the impact of material composition and fabrication methods on membrane performance and selectivity. Additionally, the chapter scrutinizes the diverse materials utilized in membrane technology, including polymeric, ceramic, and composite membranes, and their respective advantages and limitations. Moreover, it critically evaluates various membrane preparation techniques, ranging from phase inversion to 3D printing, highlighting their influence on membrane morphology and performance. Finally, the chapter concludes with a profound analysis of practical applications and the future of membranes, outlining the emerging challenges and environmental prospects in the field and advocating for innovative solutions to advance sustainable membrane technologies. This chapter serves as a valuable resource for researchers and engineers seeking to comprehend the current landscape and future directions of membrane technology.

Keywords: membrane technology, desalination, separation, challenges, future prospects

1. Introduction

Water scarcity and the prevalence of water crises have become pressing global concerns, impacting millions of people and ecosystems worldwide. With increasing population growth, industrialization, urbanization, and climate change, traditional freshwater sources are under immense pressure, leading to shortages and deteriorating water quality. In this context, desalination emerges as a vital and optimum solution to address water scarcity, particularly in arid regions and coastal areas where access to fresh water is limited.

Desalination, the process of removing salts and impurities from seawater or brackish water to produce potable water, has become increasingly important with the growing global demand for fresh water [1]. Membrane technology has emerged as a leading method for desalination, offering a sustainable and energy-efficient solution compared to traditional thermal processes such as distillation. This review aims to explore the advancements in membrane technology for desalination processes, focusing on the latest innovations, challenges, and future prospects in this rapidly evolving field [2].

Membrane separation processes have become the leading separation technology over the past two decades [3]. The membrane is defined as a selective thin layer of a semipermeable material that acts as a selective barrier and separates undesired species from a feed solution based on their sizes or affinity by exerting a potential gradient, such as pressure, temperature, electrical, or concentration difference [4, 5] as shown in **Figure 1**. Membrane-based desalination processes, including reverse osmosis (RO) and nanofiltration (NF), have shown tremendous potential in addressing the world's water scarcity issues. These processes rely on semipermeable membranes to separate salts and other contaminants from water, yielding high-quality freshwater. In recent years, significant research and development efforts have been directed toward enhancing the efficiency, productivity, and sustainability of membrane-based desalination technologies [6].

Membranes are often categorized according to their materials, shape, driving force, and average pore size. One of the most important aspects of the membrane's initial differentiation is the pore size of the substance or surface [7]. However, both organic and inorganic materials can be used as membranes [8].

The two fundamental technological approaches to desalinating water are the thermal approach and the membrane approach. The thermal approach typically involves the use of heat to evaporate water, leaving the salts and impurities behind, and then condensing the evaporated water to obtain freshwater. This method has been historically significant and is utilized in large-scale operations, often requiring substantial energy inputs. However, advancements in thermal desalination processes, such as multi-stage flash distillation and multi-effect distillation, have improved energy efficiency and reliability [9, 10].

On the other hand, membrane desalination involves the use of semi-permeable membranes to selectively separate salts and impurities from water, allowing only freshwater to pass through. This method is prevalent in modern desalination plants and is primarily carried out through reverse osmosis process. Membrane desalination

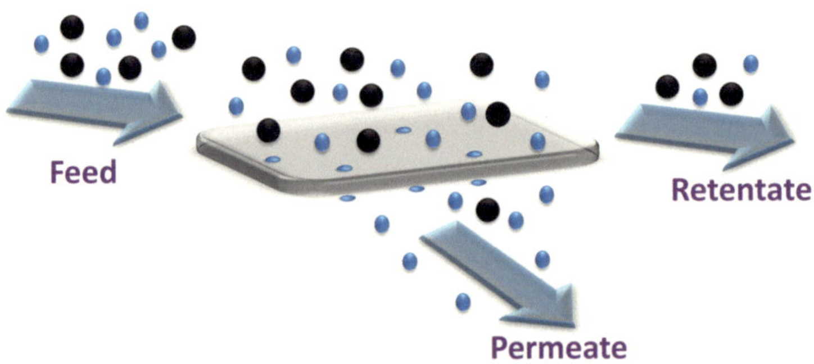

Figure 1.
Normal procedure for membrane separation.

is often preferred for its relatively lower energy requirements and modular scalability, making it suitable for diverse applications, ranging from small-scale community water treatment to large industrial facilities [11]. Each approach possesses its unique advantages and limitations, and the choice between thermal and membrane desalination methods depends on factors such as feed water salinity, energy availability, environmental considerations, and economic feasibility.

Comparing renewable energy to traditional systems reveals various advantages. For example, it can lessen reliance on fossil fuels, generate electricity with zero or very low greenhouse gas emissions, help stabilize energy prices, and create jobs and economic opportunities in the expanding clean energy sector. It is also sustainable and does not deplete natural resources [12, 13]. Membrane-based desalination systems may be powered by renewable energy sources in many ways. Pumps and other desalination system equipment may be powered directly by electricity produced from renewable energy sources like solar and wind. Membrane-based desalination systems can become more sustainable and have a lower carbon footprint if they are powered by renewable energy [14]. Additionally, it can lessen reliance on fossil fuels and help save desalination costs over time.

This chapter serves as a comprehensive guide for researchers, engineers, and industry professionals seeking to comprehend the multidimensional domains of membrane technology, enabling them to navigate the evolving landscape of membrane science and engineering.

2. Membrane separation processes

Dissolved or suspended particulates in a liquid can be effectively and economically separated via membrane filtering and desalination. Depending on their chemical and/or physical characteristics, certain chemicals can flow through the membrane, which is a physical barrier [15]. Typically, a thin, dense layer creates the real membrane on the top of a porous support layer, which is what makes up a membrane [16]. The following describes several types of membrane filtration according to membrane pore sizes.

2.1 Microfiltration

Microfiltration (MF) is a low-pressure method of separation using highly permeable membranes in the approximate size range of 0.1 to 20 μm [17]. Microfiltration is the process of passing fluids through a microporous material, such as a membrane filter, to remove particles or biological entities greater than 1000 nm [18] as shown in **Figure 2**. While non-membrane or depth materials, like those in fibrous media, can remove micron-sized particles, only a membrane filter with precisely determined pore size can guarantee quantitative retention. Microfiltration can remove impurities from water without harsh or costly chemicals because it is a physical separation method. Because of its bigger pore size, it works well for "first pass" filtration, which filters water partially before transferring it to a process like reverse osmosis or ultrafiltration [19]. It is helpful in other applications that do not call for fine filtration and can be used in the food industry to clarify liquids, such as fruit juice. The inability of microfiltration to separate microscopic items from water due to its enormous pore size is its main drawback [20].

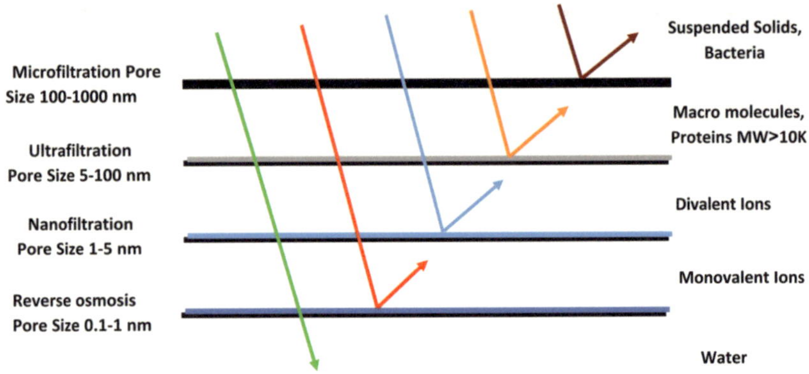

Figure 2.
Types of membrane separation processes.

2.2 Ultrafiltration

The technique known as ultrafiltration (UF) is used to extract dissolved molecules and tiny particles from liquids. Although in all filtration applications, the permeability of a filter medium can be altered by the chemical, molecular, or electrostatic properties of the sample, molecule size serves as the major basis for separation [21]. The molecules that ultrafiltration can separate are limited to different sizes.

Certain ultrafiltration membranes retain materials with molecular weights (MW) over 10 K as shown in **Figure 2**, allowing water and salts to pass through. Particulate and colloidal materials can both be held in reserve [22]. Membranes for ultrafiltration can be used to collect material that the filter has retained and purify material that passes through it. The filter lets through materials that are noticeably smaller than the pore size rating [23].

The filter retains materials bigger than the pore size rating, allowing them to be concentrated or separated from pollutants with low molecular weight. Proteins are usually separated from buffer components via ultrafiltration in preparation for buffer exchange, desalting, or concentration. Additionally, ultra-filters are perfect for exchanging or removing sugars, non-aqueous solvents, removing compounds with a low molecular weight, separating free of protein-bound ligands, and quickly changing the pH and/or ionic environment in order to retain specific proteins [24].

Compared to procedures like precipitation, ultrafiltration is significantly softer to solutes. Because it can simultaneously concentrate and desalt solutes, UF is more effective. Phase changes, which frequently denature labile species, are not necessary for UF, and it can be carried out in a cold environment or at room temperature. On the other hands, it may not effectively remove low molecular-weight solutes or ions, and it requires regular cleaning to prevent fouling [25, 26].

2.3 Nanofiltration

Nanofiltration (NF) is a membrane separation process that falls between ultra-filtration and reverse osmosis in terms of the size of particles it can effectively separate. The membranes used in nanofiltration have smaller pore sizes compared to ultrafiltration as depicted in **Figure 2**, allowing them to selectively remove divalent ions, organic matter, and smaller particles while partially retaining monovalent ions [27]. This makes nanofiltration particularly useful for applications such as water softening, color removal, and partial desalination [22].

- *Advantages of NF*:
    - Selective Separation: It can effectively remove divalent ions while retaining monovalent ions, leading to selective separation and purification of specific components in a liquid [28].
    - Reduced Energy Consumption: Nanofiltration typically requires lower operating pressures compared to reverse osmosis, resulting in reduced energy consumption [29].
    - Versatility: It can be tailored to various applications, including water and wastewater treatment, as well as in industries such as food and beverage, pharmaceuticals, and chemical processing [30].
- *Limitations of NF*:
    - Sensitivity to Fouling: Membranes used in nanofiltration are susceptible to fouling and scaling, necessitating careful pretreatment of feed water to prevent membrane damage and maintain performance [31].
    - Operating Costs: While it requires lower pressures than reverse osmosis, nanofiltration still involves energy and operational costs that must be considered [32].
    - Variable Performance: The performance of nanofiltration can vary based on the specific compositions of the feed solution, which may require careful process optimization [33].
    - Nanofiltration has found widespread use in both industrial and environmental applications due to its ability to selectively separate and purify components in liquids, but its effective application requires careful consideration of the specific properties and challenges of the feed solution [22].

2.4 Reverse osmosis

Reverse osmosis (RO) is a water purification process that uses a partially permeable membrane to remove ions, molecules, and larger particles from water as its pore sizes range from 0.1 to 1 nm [6] as shown in **Figure 2**. It is a highly effective method for producing clean drinking water and is widely used in both residential and industrial settings. It works through two dependently methods, filtration – where water is passed through a semi-permeable membrane, which allows water molecules to pass through while trapping contaminants such as ions, molecules, and particles – and osmotic pressure – where in reverse osmosis, a pressure greater than the osmotic pressure is applied to the water, forcing it through the membrane, leaving behind contaminants [34].

- *Advantages of RO*:
  - o High Purity: RO systems can remove a wide range of impurities, including dissolved salts, minerals, and organic compounds, resulting in high-purity water [35].
  - o Versatility: RO systems are used in various applications, including desalination of seawater, production of drinking water, and purification of industrial process water [36].
  - o Ease of Use: RO systems can be designed for both large-scale industrial use and small-scale residential applications, providing clean water for different purposes [34].
- *Limitations of RO*:
  - o Wastewater Generation: RO systems produce concentrated wastewater (brine) as a byproduct, which requires proper disposal or treatment [37].
  - o Energy Intensive: The process of applying pressure to overcome osmotic pressure requires energy, making RO systems relatively energy-intensive compared to other water treatment methods [11].
  - o Membrane Sensitivity: The effectiveness of RO systems can be impacted by fouling, scaling, and damage to the semi-permeable membranes, requiring regular maintenance and cleaning [38].

Reverse osmosis is an efficient method for producing high-purity water and has a wide range of applications, but it is important to consider the energy requirements, wastewater management, and maintenance needs associated with its use.

3. Membrane structure and synthesis techniques

Membranes can be categorized according to their applications, separation mode, manufacturing process, geometry, and general structure [39]. **Figure 3** depicts this basic classification structure. Membranes can be made as hollow fiber tubes or as flat sheets. Tubular membranes are packaged in hollow fiber modules; flat sheet membranes are packaged in sheet and frame modules or spiral wrapped modules. Large-scale separation operations also frequently use spiral coil and plate-and-frame modules, despite hollow fiber modules having the maximum membrane packing density per module volume. The structure of membranes can be either asymmetric (anisotropic) or symmetric (isotropic) [40]. Asymmetric membranes have a deteriorated structure, whereas symmetric membranes have a consistent structure across their whole thickness. Symmetric membranes' general structure dictates their separation characteristics. Conversely, the densest part of an asymmetric membrane mostly determines its separation characteristics [41]. **Figure 4** displays the most prevalent varieties of symmetric and asymmetric membranes.

3.1 Symmetric membranes

Membranes that are symmetrical have the same content and structure over their whole thickness. On both sides, their permeability and selectivity are the same.

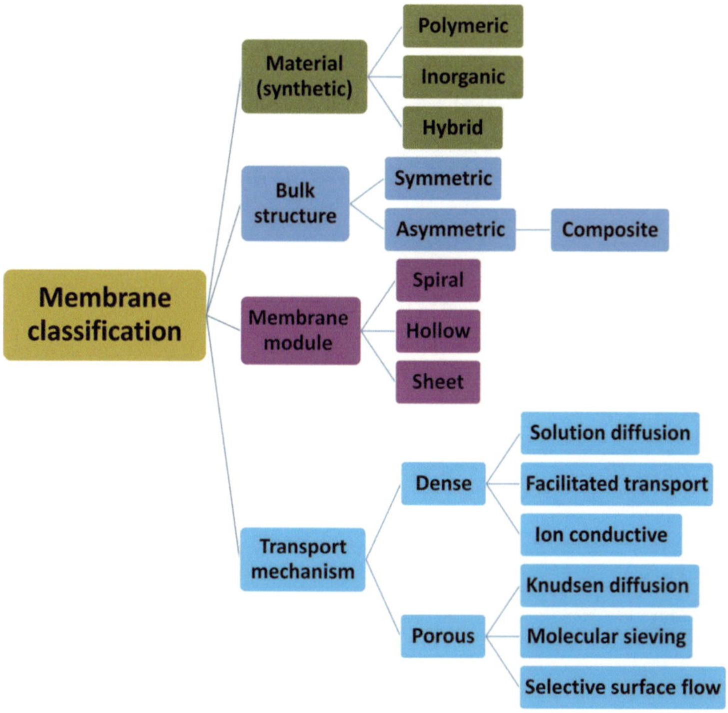

Figure 3.
Overview of membrane classification.

These membrane types are frequently employed in situations where the components to be separated must be separated based on their molecular size or shape [43].

- *Porous Symmetric Membranes*: Symmetric porous membranes can be made in a number of ways and typically feature structures structured like sponges, slits, strips, or cylindrical shapes. The most crucial techniques for creating symmetric porous films are (i) radiation; (ii) stretching melt-treated semi-crystalline polymer films; (iii) phase separation caused by vapor; and (iv) temperature-induced phase separation [44]. Melt extrusion/stretching can be used to create membranes with a symmetric slit-shaped porous structure from semicrystalline polymers like polyethylene and polypropylene. Initially, a semicrystalline polymer is melted extruded and then recrystallized under high pressure to generate a row-nucleated lamellar structure. The film is stretched in the machine's direction in the second phase to create slit-like perforations between the stacked sheets [45].

By using a vapor deposition/evaporation technique, symmetric membranes with pore architectures resembling sponges or webs can be created (6). This method involves pouring solutions containing polymers, solvents, and non-solvents onto an appropriate substrate and exposing it to a stream of air saturated with water vapor. The originally stable polymer solution undergoes phase separation due to steam [46]. A heated air stream is sent across the membrane to evaporate the solvent and non-solvent components following phase separation. Vapor precipitation/evaporation produces a typical membrane cross-section. This kind of membrane's porosity and

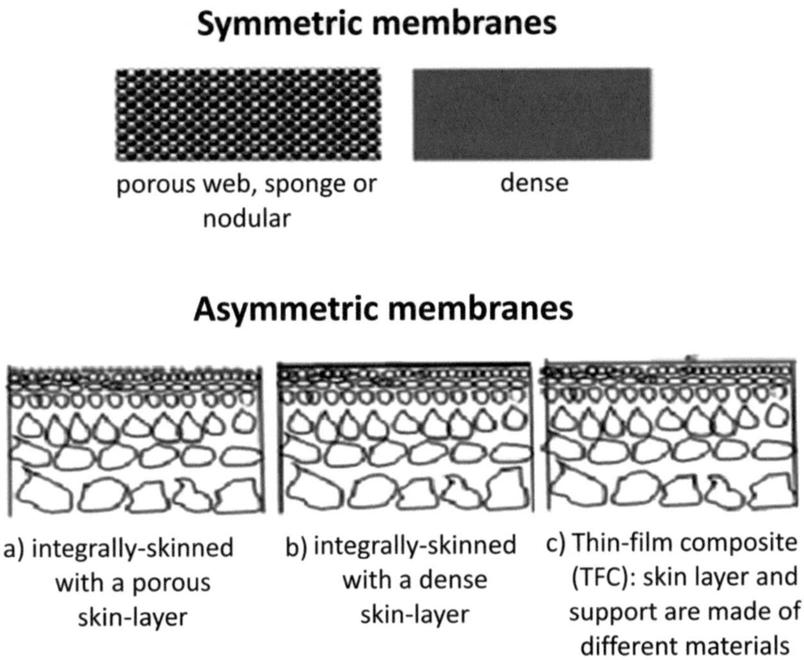

Figure 4.
Schematic of (top) symmetric membranes (porous symmetric and dense symmetric); (bottom) asymmetric membranes (a) porous skin layer on a porous sublayer; (b) selective dense skin layer on a porous sublayer; (c) selective dense coating layer on a porous sublayer) [42].

pore size can be regulated by (i) the amount of polymer present in the casting solution and (ii) the composition of the vapor atmosphere. Films with high porosity and large pore sizes are produced by adding solvent vapor to the molding media, reducing the concentration of polymer, and increasing humidity [47].

- *Dense Symmetric Membranes:* Dense symmetric films can be made by melt extrusion or solution casting with solvent evaporation afterward, with thicknesses larger than 10 μm. The permeability and selectivity of the novel membrane material were evaluated using this kind of membrane to determine its material-specific transport characteristics. These membranes have a limited usage in the air separation process due to their relative thickness [48]. But in electrolysis applications, dense symmetric membranes are frequently employed.

3.2 Asymmetric membranes

Asymmetric membranes, on the opposite hand, have a layered structure with discrete layers that differ in composition or pore size. They have a porous support layer offering mechanical strength and a dense active layer in charge of the separation process. When it comes to selectivity and flow, asymmetric membranes are superior to symmetric membranes [49]. They are frequently employed in several filtration and separation procedures, including gas separation and water treatment.

Asymmetric membranes can be divided into three main types as shown in **Figure 4**: thin-film composite (TFC) membranes, asymmetric integrals with porous skin, and asymmetric integrals with dense skin. Denser integrated asymmetric membranes provide better selectivity and flow than symmetric membranes [40]. Integrated asymmetric porous membranes are produced using a phase inversion method and used in dialysis, ultrafiltration, and microfiltration applications. They are frequently employed in a variety of filtration and separation procedures, including gas separation and water treatment [50].

- *Integrally-Skinned Asymmetric Membranes:* In the early 1960s, Loeb and Sourirajan designed the first asymmetric, integrally-skinned cellulose acetate membranes for water desalination [51]. Membrane-based desalination became economically feasible very rapidly because these membranes were significantly more effective at desalinating than any other membrane available at the time.

The common method for synthesizing these asymmetric membranes is immersion precipitation. Integrally-skinned asymmetric membranes are made using a binary solution made of a polymer and a solvent. When the cast solution is submerged in a liquid that is miscible with the solvent but not a solvent for the polymer, an asymmetric structure with either a porous or non-porous skin layer forms [52]. The structural gradient in integrally skinned asymmetric membranes is caused by the extremely steep polymer concentration gradient in the growing membrane at the start of phase separation [50]. Typically, the immersion precipitation membrane is composed of a highly porous substructure and a thin skin layer.

- *Thin-Film Composite Membranes:* At least two structural parts made of different materials form the composite membranes. The two fundamental types of thin-film composite membranes are schematically shown in **Figure 5**. A single-layer composite membrane consisting of a thin, selective top layer and a porous support is shown in **Figure 5a**. The thin top layer manages separation, while the porous support provides mechanical strength. A multi-layer composite membrane with multiple layers of different materials, each with a specific function [54], is shown in **Figure 5b**. The membrane is composed of a porous substrate. Thin-film

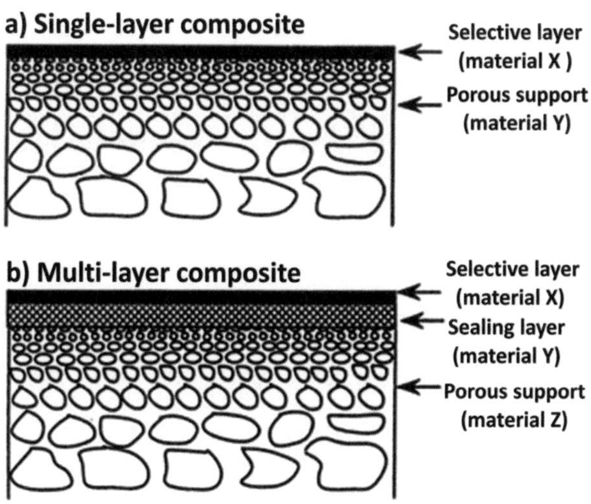

Figure 5.
Schematic for thin-film composite membranes: Single-layer thin-film composite membranes (a) and multi-layer membranes (b) [53].

composite membranes are generally employed in processes where permeation is controlled by the solution-diffusion mechanism, such as nanofiltration, reverse osmosis, gas separation, and pervaporation. The selective layer can be applied using lamination, solution coating, interfacial polymerization, or plasma polymerization processes [55].

4. Membranes materials

Membrane materials, the fundamental component of the membrane separation process, significantly impact the membrane's performance. The three primary types of membranes are inorganic, organic, and hybrid [56]. Hybrid membranes are composed of both inorganic and organic components. Membranes are used in various sectors, including gas separation, water treatment, and membrane-based sensors. Each form of membrane has unique characteristics of its own.

4.1 Inorganic membranes

Membranes can also be produced from inorganic materials such as glass, metal oxides (such as titanium dioxide and zeolites), and ceramics (such as silica, alumina, and zirconia) [57]. Membranes can be broadly classified into two types: dense (composed of mixed conductive oxides, metals, or hybrid organic-inorganic materials) and porous (ceramic) [57]. Sol-gel, hydrothermal synthesis, and plasma-enhanced chemical vapor deposition are methods for producing inorganic membranes [58]. Inorganic and organic membranes compete for specific uses in harsh environments. Many ceramic membranes are functional at temperatures over 1000°C, and metal membranes remain stable at temperatures between 500 and 800°C [59]. As a result, they can operate at high temperatures. They are significantly more resistant to chemicals and have a longer lifespan. Their cost, ability to modify the surface, and pore properties, however, could not make them competitive. Consequently, affinity membranes are seldom supported by inorganic materials.

4.2 Organic membranes

Membranes made from organic materials are increasingly gaining attention for their applications in filtration and separation processes. These membranes, often derived from natural polymers like cellulose, chitosan, and protein-based substances, offer several advantages, including biodegradability, cost-effectiveness, and tunable properties [60]. Their porous structures can be engineered to selectively filter out pollutants, microorganisms, and contaminants from water or air, making them ideal for environmental remediation and safe drinking water production [61]. Additionally, organic membranes can be designed to allow for specific separation processes in chemical industries, enhancing the efficiency of resource recovery and waste management. With ongoing research, the performance and durability of these organic membranes are continually improving, paving the way for sustainable filtration technologies. The use of renewable materials not only reduces dependency on fossil fuels but also minimizes environmental impact. Biodegradable membranes can break down naturally, reducing landfill waste compared to synthetic alternatives [62]. While promising, organic membranes face challenges such as lower mechanical stability and sensitivity to environmental conditions. Ongoing research is focused on enhancing their durability, scalability, and performance through hybrid approaches (combining organic and inorganic materials) and advanced engineering solutions.

The process for fabricating polymer membranes is determined by the polymer used and the membrane's intended structure. The most prevalent methods for preparing polymeric membranes include interfacial polymerization, phase inversion, dip coating, stretching, track-etching, electrospinning, template-leaching, and 3D printing as shown in **Figure 6**.

4.3 Interfacial polymerization

Interfacial polymerization (IP) is a technique used to create polymers at the interface of two immiscible phases, typically a liquid-liquid or a liquid-gas interface. This process involves the reaction between two or more monomers at the interface, leading to the formation of a polymer film [63].

In a typical IP method, an aqueous diamine (e.g., m-phenylenediamine, MPD) solution is used to pre-wet a porous substrate, which is then placed in contact with the organic phase of acyl chloride (e.g., trimesoyl chloride, TMC) [64]. The resulting thin polyamide layer at the interface of aqueous and organic phases enables fast-selective mass transport for simple membrane separation, while the porous substrate provides

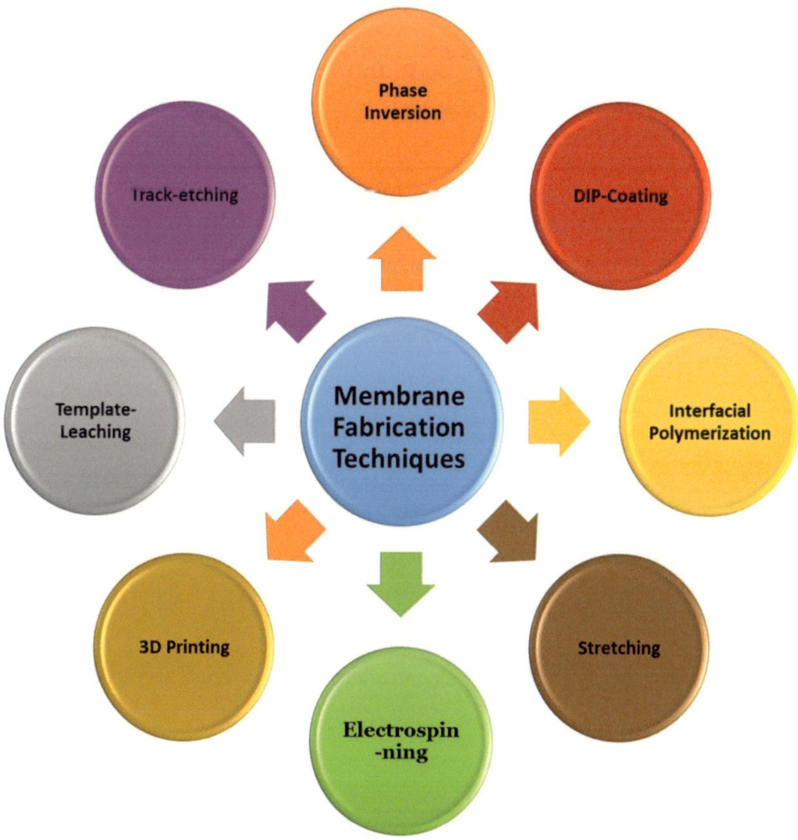

Figure 6.
Membrane fabrication techniques.

mechanical strength to meet the demands of pressurized filtration, resulting in synergistic effects that maximize separation efficiency in real-world applications [65]. Currently, TFC membranes are among the top technologies in RO-based desalination. Because effective separation occurs in the thin layer, researchers have been trying to examine the impact of polymer chemistry on IP through synthesis of monomers and including additives. Porous substrates, which serve as a support for thin layers atop TFC membranes, have received far less attention [66].

4.4 Phase inversion

Originally invented by Loeb and Sourirajan for cellulose acetate desalination membranes as mentioned before, it eventually became popular in the manufacture of polymeric membranes for gas separation because of its simplicity, economic practicality, and capacity to produce thin membranes [67]. To produce a viscous polymer dope, a water-insoluble polymer is dissolved in a solvent with a high boiling point. This solution is then degassed and placed onto a support with a casting rod or knife before being immersed in a coagulation bath of water or solvent, yielding an asymmetric membrane with a dense top layer and a porous bottom layer [68]. The solution composition, coagulation temperature, and additives all impact the creation of these separate layers as illustrated in **Figure 7**.

There are three primary types of phase inversion techniques: thermal induced phase inversion (TIPS), non-solvent induced phase inversion (NIPS), and vapor-induced phase inversion (VIPS). Despite the ease and widespread usage of phase inversion and solvent evaporation methods, the numerous obstacles connected with them highlight the critical need for careful study and new ways that will lead future research and development [69].

4.5 Dip-coating

In order to create a thin layer on the surface, dip coating entails submerging a substrate in a coating solution and pulling it out quickly [70]. One of the most basic methods of film deposition, it is especially useful for making gas separation

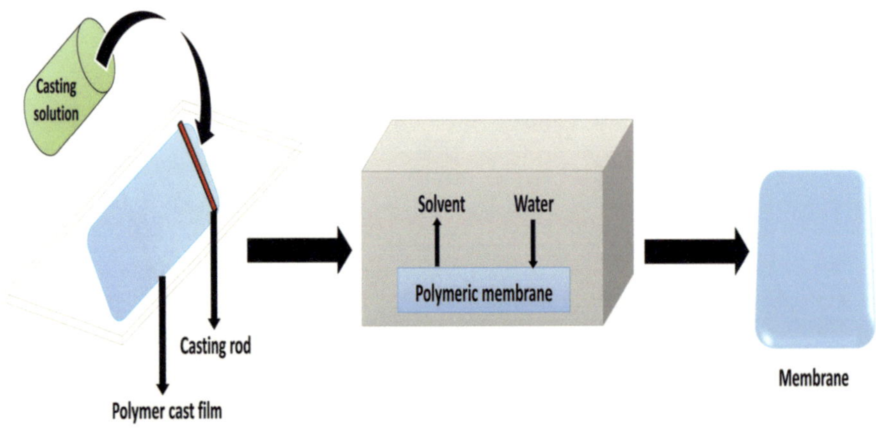

Figure 7.
Preparation of polymeric membranes through casting phase inversion technique.

membranes because it makes it easier to create the thin layers that need to be regular and exact in order for the membrane to function as best as possible [71].

The process works well for creating membranes for the separation of different gases, such as O_2/N_2 [72], CO_2/CH_4 [73], and CO_2/N_2 [74]. It was determined that the ideal temperature and casting solution concentration were crucial for producing a homogeneous membrane. Studies have indicated that variables like coating solution concentration, withdrawal rate, and quantity of coatings applied are critical in adjusting the gas separation properties of the membrane, emphasizing the significance of process control in attaining the intended membrane functionality [69]. Moreover, the process of dip-coating has demonstrated remarkable efficacy in producing flawless composite membranes for gas separation, highlighting the versatility and effectiveness of the approach. Nevertheless, there are a number of drawbacks to this method that may restrict its usefulness and make scaling more difficult [75]. Initially, because the solution tends to cover the whole submerged surface, this approach might not be the best choice for applications that simply require coating on one side of a substrate. It also presents a lot of complications since a large volume of precursor solution is needed to fill the coating container [76]. For example, handling large amounts of the solution might be dangerous and negatively affect the environment; these limitations impede scaling.

4.6 Stretching

Porous hollow fiber membranes can be made using either of the two methods, melt-stretching or TIPS [77]. Phase separation is the next step after casting the polymer solution in TIPS to manufacture the polymer membrane. On the other hand, melt-stretching is a method of preparation that involves uniaxially stretching a semi-crystalline polymer film that has a lamellar structure with rows nucleated. Because melt-stretching does not require a solvent, it is more cost-effective and ecologically benign than TIPS [78].

This method comprises three main sequential processes: (1) melt-extrusion is used to produce a precursor film with a lamellar structure [77]; (2) the precursor film is annealed at temperatures close to the polymer's melting point to remove defects in the crystalline phase, improving the crystalline structure and making the lamellae thicker and more uniform [79]; and (3) the precursor film is stretched at low temperature to create voids and then at high temperature to enlarge the pores. The structure and functionality of the porous polymeric membrane are determined by the parameters of each process, thus choosing the right parameter for each process is crucial [79].

4.7 Electrospinning

Electrospinning is a widely used technique for fabricating nanofibrous membranes, renowned for its ability to produce fibers with diameters in the nanometer range [80]. The process begins by dissolving a polymer (such as polycaprolactone, polylactic acid, or various biopolymers) in a suitable solvent to create a viscous spinning solution [81]. This solution is loaded into a syringe connected to a metal nozzle and subjected to a high-voltage electric field typically ranging from 5 to 30 kV or more [82]. As the electric field is applied, the polymer solution is drawn out of the nozzle, forming a thin jet that stretches and dries as it travels toward a collector plate, typically held at ground potential [80] as shown in **Figure 8**. As the solvent evaporates, ultrafine fibers are deposited on the collector, forming a nonwoven

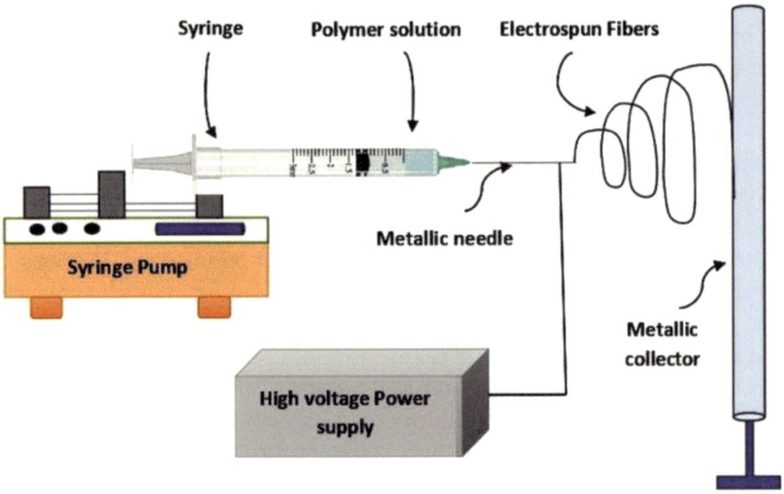

Figure 8.
Electospinning basic set up [83].

mat or membrane. The fiber diameter, orientation, and porosity can be tailored by adjusting parameters such as the solution concentration, flow rate, voltage, and collector distance. Optimal conditions must be carefully determined, as slight changes in parameters can significantly affect fiber morphology. Due to their high surface area-to-volume ratio, porosity, and biocompatibility, electrospun membranes find applications in areas such as tissue engineering, drug delivery, filtration, and protective clothing [84].

4.8 Track-etching

Track etching is a specialized technique used for fabricating membranes with high precision and uniform pore structures [85]. This method involves irradiating a polymer film, typically made of materials like polycarbonate or polyethylene terephthalate, with high-energy particles (such as heavy ions) [86]. The radiation creates latent tracks—damaged regions—in the polymer. Following irradiation, the membrane is subjected to a selective chemical etching process, which removes the damaged areas, resulting in cylindrical pores that penetrate through the thickness of the film. By controlling the etching conditions, such as time and temperature, the pore size and density can be finely tuned, making track-etched membranes ideal for applications in filtration, biomolecule separation, and microfluidics [87]. Overall, this technique offers outstanding reproducibility and versatility, allowing for the production of membranes with tailored properties for diverse scientific and industrial applications. The process has several drawbacks, including the requirement for specialist equipment for the etching and radiation processes and potential restrictions on the kinds of materials that may be utilized based on their compatibility with the etching solution [88].

4.9 Template-leaching

The method known as template-leaching can be used to create porous membranes from polymer materials with a variety of internal and exterior geometries [89].

For example, it has been applied to create silica glass membranes with pore sizes as tiny as 0.05 μm [90]. This method involves combining a leachable component with a membrane matrix material to create a homogenous film. A macromolecule like polyethylene glycol (PEG) or poly vinyl alcohol (PVA) can also be a soluble low molecular weight compound that is leachable. Following the film's preparation, the leachable component is eliminated using an appropriate chemical treatment. The size and structure of the separated phase domains in the composite may be altered by adjusting the formation factors (such as composition, temperature, and blending technique) [90]. This leads to the availability of porous matrices with a broad range of porosity and structure. Using this technique, polymeric microporous membranes may be made quickly.

4.10 3D printing

3D printing has emerged as an innovative technique for membrane fabrication, allowing for the production of highly customized and complex membrane architectures that traditional methods may not easily achieve. This additive manufacturing process involves depositing material layer by layer to create a three-dimensional structure, enabling precise control over pore size, distribution, and overall geometry of the membrane [91]. Commonly used materials include polymers, ceramics, and metal-organic frameworks, which can be tailored to meet specific requirements for various applications such as filtration, water purification, and gas separation. The integration of digital design tools facilitates rapid prototyping and the ability to fabricate membranes with intricate patterns that enhance permeability and selectivity [92]. Additionally, 3D printing technologies can enable the incorporation of functional components, such as sensing elements, directly into the membrane structure, paving the way for advanced applications in smart filtration systems and bioreactors. Despite restrictions in cost, speed, printing resolution, and material selection, 3D printing has made significant advances in membrane module design, composite membrane fabrication, the creation of oil-water separation and wastewater treatment materials, and so on [93]. A growing number of research organizations are using 3D printing to create complex structures with ease of prototyping and testing for diverse water-related applications.

5. Practical applications of membrane technology

Membrane technology offers a wide array of practical applications across various industries, significantly enhancing efficiency and sustainability [2]. In water treatment, membranes are employed for desalination and wastewater purification, allowing for the effective removal of contaminants and the recovery of valuable resources, such as nutrients and water [94]. In the food and beverage sector, membranes are used for processes like microfiltration and ultrafiltration, improving product quality while minimizing energy consumption and chemical use [95]. Additionally, in the pharmaceutical industry, membrane technology is crucial for drug purification and concentration, ensuring high purity levels critical for patient safety [84]. In the industrial processes, in the textile industry for instance, membranes can help recover dyes and chemicals from wastewater, allowing for their reuse in production processes [96]. This not only cuts costs but also significantly reduces environmental pollution associated with dye discharge. Furthermore, membranes play a vital role in gas separation, helping capture carbon dioxide from industrial emissions and contributing to climate change mitigation efforts [97]. The benefits of membrane technology include reduced environmental impact, lower operational costs, and the ability to create more efficient production processes, making it a key component of sustainable practices in modern industries.

Membrane technology plays a crucial role across various sectors by enhancing efficiency, reducing waste, and promoting sustainable practices. Its ability to separate and purify materials makes it indispensable in tackling modern challenges like water scarcity, environmental pollution, and energy needs.

6. Membranes desalination technologies

There are two basic technical ways to desalinate water: thermal and membrane systems, as illustrated in **Figure 9**. Several methods have been used in the desalination process throughout the years, including reverse osmosis (RO), nanofiltration (NF), multiple-effect distillation (MED), mechanical vapor compression (MVC), multi stage flashing (MSF), and membrane distillation (MD) [98]. Thermal desalination is the process of evaporating saltwater using heat energy, leaving only clean water after condensation. Thermal desalination systems are widely used in the Middle East because to the availability of fossil fuels, low-quality saltwater, and suitability for power cogeneration [99]. Even while thermal techniques produce more water with total dissolved solids that vary between 5 and 50 ppm, the system has the disadvantage of requiring a substantial amount of energy to complete the removal process, which results in high operational costs [100].

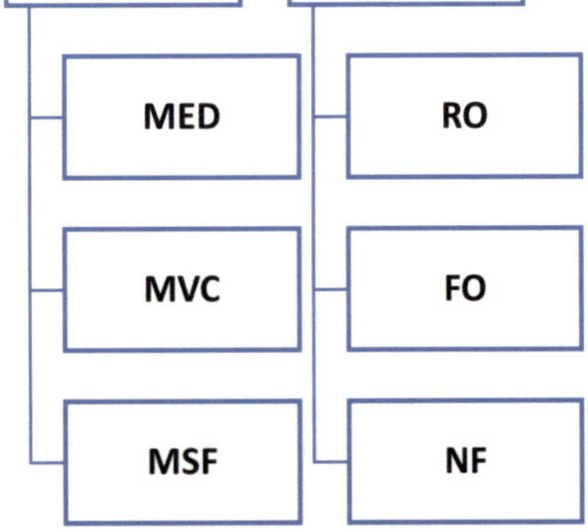

Figure 9.
Classification of desalination technologies.

When it comes to membrane technologies, which employ semi-permeable membranes to remove solutes from water, reverse osmosis is the most prevalent. RO uses strong hydraulic pressure to overcome the water's osmotic pressure as shown in **Figure 10**. The tiny pore diameters of semi-permeable membranes enable the (H_2O) solvent to flow freely but block the passage of solutes trapped on the membrane's pressure side [11]. The RO process produces purified water and a highly salted solution. The recovery rate (percentage) is the ratio between the end desalinated water and the initial water input, and it is frequently more than 60% for commercial systems [101].

Forward osmosis (FO) provides an osmotic pressure differential across a semipermeable membrane to separate different concentrations of solutions for water penetration [102]. Water can cross a semipermeable membrane from the feed solution to the concentrated solution (draw solution) as shown in **Figure 10** when two distinct solution concentrations are maintained on its two sides due to an osmotic pressure differential. Both the feed solution and the draw solution are aimed at the membrane active layer and support layer, respectively. The draw solutions are located in the opposite direction as pressure-retarded osmosis. One example that operates at a lower pressure than RO and requires less energy is FO [102]. Membrane distillation (MD) is a technique that employs microporous membranes and heat separation to treat highly salinized water; however, this technology was not employed extensively until the 1960s [9].

More energy is required for thermal-based processes than for membrane-based ones. Salt concentration has little effect on the energy requirements for thermal desalination systems; however, salt content influences the requirements for membrane processes. Perhaps, the most well-known advantage of membrane technologies over thermal ones is how much less energy they require. RO energy needs stay at 3 to 4 (kWh/m^3) for saltwater or drop to 0.5 to 2.5 (kWh/m^3) for brackish water, whereas MSF requires between 10 and 16 (kWh/m^3) and MED uses between 5.5 and 9 (kWh/m^3) [99].

Water passes through a semipermeable membrane during the desalination process when pressure is applied to the saltwater feed solution that is greater than either its osmotic pressure or the minimum pressure that prevents the intake of clean water by osmosis [103]. Because RO increases osmotic pressure and accelerates the passage of salt over the membrane, it is not recommended for desalinating extremely concentrated salt solutions [34]. Working with very salty solutions is better suited for membrane distillation, an alternative desalination technique that combines heat and

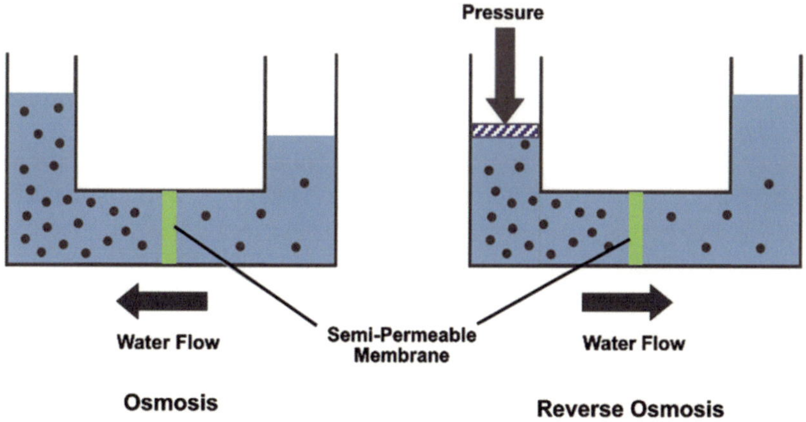

Figure 10.
Schematics of osmosis and reverse osmosis process [1].

membrane processes. This is especially true for solutions with salinities between 70 and 300 g of salt per kilogram of solution [98].

7. The future of desalination membranes: challenges and environmental prospects

Current and future challenges in membrane desalination pose significant opportunities for innovation and sustainable development. Addressing the pressing need for greater energy efficiency remains a key challenge, with a focus on reducing the operational costs and environmental impact of desalination processes [104]. Additionally, combating membrane fouling through the development of anti-fouling membranes and effective cleaning techniques is essential for maintaining the long-term performance and viability of desalination systems. Moreover, as desalination capacity grows, the environmental impact of concentrated brine disposal necessitates careful consideration, driving the exploration of brine management and resource recovery solutions [105–109]. The quest for cost reduction, advancements in materials for improved membrane performance, and ensuring high water quality standards further shape the landscape of challenges and opportunities in membrane desalination. Simultaneously, the integration of renewable energy sources, like solar and wind power, into desalination processes emerges as a critical future endeavor for minimizing environmental impact. Tackling these challenges will pave the way for more sustainable and accessible water supply solutions in water-stressed regions, driving innovation and progress in membrane desalination.

Over 90 million m^3/day of desalinated water may now be produced through existing desalination capacity, and the global direct carbon footprint of desalination is projected to be over 120 million metric tons per year [110, 111]. Because of the present climate issue, which calls for the sector to decarbonize, the use of renewable energy must be linked to the growing desalination process.

Membrane methods account for more than 70% of desalination globally, and they all need electricity [4]. Solar, wind, and wave energy are the most appropriate renewable energy sources for desalination since hydroelectric power and biomass both inherently require water [99]. Solar energy is most common because of the region's high solar radiation and water shortage. Similar to FO, an ascending desalination method like MD employs low temperature heat to some degree [9]. Geothermal and solar energy can provide this. Technical advancements are required to make it easier to connect renewable energy sources to membrane technology for desalination. Membrane desalination and renewable energy are quite diverse technologies; thus, knowledge of both areas is needed.

In the realm of membrane materials, the future holds exciting prospects for cutting-edge advancements. The development and integration of novel materials, such as graphene-based membranes, metal oxide frameworks (MOFs), carbon nanotubes, and nanostructured materials hold promise for revolutionizing desalination and water purification processes [112]. These materials exhibit exceptional mechanical strength, chemical resistance, and high permeability, offering the potential for enhanced water flux and improved separation efficiency. Additionally, the emergence of responsive or "smart" membrane materials, capable of dynamically adjusting their permeability in response to external stimuli, opens new avenues for precise and energy-efficient water treatment processes [113]. Furthermore, the exploration of sustainable and biodegradable membrane materials aligns with the growing emphasis on eco-friendly technologies, aiming to reduce environmental impact and promote circular economy principles within the membrane industry [114]. As research and development efforts continue, the integration of nanotechnology and advanced manufacturing techniques holds the potential to unlock unprecedented levels of performance and durability in membrane materials. These innovative strides are poised to propel the field of membrane materials toward a future defined by heightened efficiency, sustainability, and adaptability, shaping the landscape of water treatment and resource management around the globe.

8. Conclusion

In conclusion, the chapter on membrane technology has navigated through the intricate realms of membrane separation processes, membrane structure and synthesis techniques, membrane materials, membrane preparation techniques, and the future of membranes, culminating in a profound understanding of the current landscape and the potential trajectories that lie ahead.

Throughout this exploration, we have underscored the pivotal role of membrane technology in addressing pressing challenges related to resource sustainability, environmental preservation, and industrial efficiency. By elucidating the fundamental principles that underpin membrane separation processes and delving into the intricate interplay between material composition, fabrication methods, and membrane functionality, we have laid a solid foundation for comprehending the diverse applications and implications of membranes across industries.

Moreover, the critical assessment of membrane materials and preparation techniques has highlighted the importance of material selection and fabrication methodologies in optimizing membrane performance, selectivity, and applicability to real-world applications and challenges. This deep understanding serves to guide researchers, engineers, and industry professionals in their pursuit of tailored membrane solutions for specific applications, thereby fostering innovation and efficiency in diverse domains.

In essence, this chapter stands as a comprehensive guide for those seeking to navigate the multifaceted landscape of membrane technology, serving as a testament to the unyielding potential and relevance of membranes in shaping a sustainable and resilient future. As we embark on this illuminating journey through the annals of membrane science and engineering, let us embrace the challenges and opportunities that lie ahead, driven by a collective vision of catalyzing positive change through the transformative power of membrane technology.

The future is ripe with possibilities, and it is within our grasp to mold the destiny of membranes into a force for sustainable progress and environmental conservation.

Author details

Mahmoud Shaban[1*], Ashraf Morsy[2,3], Amira S. Mahmoud[4] and Hamida Abdel-Hamid[1]

1 Faculty of Science, Department of Chemistry, Alexandria University, Egypt

2 Department of Materials Science, Institute of Graduate Studies and Research, Alexandria University, Egypt

3 Faculty of Engineering, Department of Petrochemical Engineering, Pharos University, Alexandria, Egypt

4 Department of Environment Studies, Institute of Graduate Studies and Research, Alexandria University, Egypt

*Address all correspondence to: mahmoud.shabaan@alexu.edu.eg

References

[1] Widiasa IN, Paramita V, Kusumayanti H. Bwro desalination for potable water supply. Journal of Coastal Development. 2009;**12**(2):81-88

[2] Osman AI et al. Membrane technology for energy saving: Principles, techniques, applications, challenges, and prospects. Advanced Energy and Sustainability Research. 2024;**5**(5):1-29. 2400011. DOI: 10.1002/aesr.202400011

[3] Ma R et al. The application of membrane separation technology in the pharmaceutical industry. Membranes (Basel). 2024;**14**(1):1-13. DOI: 10.3390/membranes14010024

[4] Politano A et al. 2024 roadmap on membrane desalination technology at the water-energy nexus. Journal of Physics: Energy. 2024;**6**(2):021502. DOI: 10.1088/2515-7655/ad2cf2

[5] Tekinalp Ö, Zimmermann P, Holdcroft S, Burheim OS, Deng L. Cation exchange membranes and process optimizations in electrodialysis for selective metal separation: A review. Membranes (Basel). 2023;**13**(6):566. DOI: 10.3390/membranes13060566

[6] Pangarkar BL, Sane MG, Guddad M. Reverse osmosis and membrane distillation for desalination of groundwater: A review. ISRN Materials Science. 2011;**2011**:1-9. DOI: 10.5402/2011/523124

[7] Xu Z, Fan Z, Shen C, Meng Q, Zhang G, Gao C. Porous composite membrane based on organic substrate for molecular sieving: Current status, opportunities and challenges. Advanced Membranes. 2022;**2**:100027. DOI: 10.1016/j.advmem.2022.100027

[8] Terbish N, Popuri SR, Lee C-H. Improved performance of organic–inorganic nanocomposite membrane for bioelectricity generation and wastewater treatment in microbial fuel cells. Fuel. 2023;**332**:126167. DOI: 10.1016/j.fuel.2022.126167

[9] Bin Abid M, Wahab RA, Salam MA, Gzara L, Moujdin IA. Desalination technologies, membrane distillation, and electrospinning, an overview. Heliyon. 2023;**9**(2):e12810. DOI: 10.1016/j.heliyon.2023.e12810

[10] Shemer H, Wald S, Semiat R. Challenges and solutions for global water scarcity. Membranes (Basel). 2023;**13**(6):612. DOI: 10.3390/membranes13060612

[11] Greenlee LF, Lawler DF, Freeman BD, Marrot B, Moulin P. Reverse osmosis desalination: Water sources, technology, and today's challenges. Water Research. 2009;**43**(9):2317-2348. DOI: 10.1016/j.watres.2009.03.010

[12] Ali A, Tufa RA, Macedonio F, Curcio E, Drioli E. Membrane technology in renewable-energy-driven desalination. Renewable and Sustainable Energy Reviews. 2018;**81**:1-21. DOI: 10.1016/j.rser.2017.07.047

[13] Tashtoush B, Alyahya W, Al Ghadi M, Al-Omari J, Morosuk T. Renewable energy integration in water desalination: State-of-the-art review and comparative analysis. Applied Energy. 2023;**352**:121950. DOI: 10.1016/j.apenergy.2023.121950

[14] Albloushi A et al. Renewable Energy-Powered Membrane Systems for Water Desalination. Amsterdam: Elsevier Inc.; 2018. DOI: 10.1016/B978-0-12-813545-7.00007-6

[15] Hsieh HP. Chapter 4 physical, chemical and surface properties of inorganic membranes. In: Inorganic Membranes for Separation and Reaction. Vol. 3. Amsterdam: Elsevier; 1996. pp. 93-147. DOI: 10.1016/S0927-5193(96)80023-4

[16] Zhang Q, Zhou R, Peng X, Li N, Dai Z. Development of support layers and their impact on the performance of thin film composite membranes (TFC) for water treatment. Polymers. 2023;**15**(15):3290. DOI: 10.3390/polym15153290

[17] Anis SF, Hashaikeh R, Hilal N. Microfiltration membrane processes: A review of research trends over the past decade. Journal of Water Process Engineering. 2019;**32**:100941. DOI: 10.1016/j.jwpe.2019.100941

[18] Jarrar R, Abbas MKG, Al-Ejji M. Environmental remediation and the efficacy of ceramic membranes in wastewater treatment—A review. Emergent Materials. 2024;**7**(1):1-33. DOI: 10.1007/s42247-024-00687-0

[19] Karmakar S, Barman SR. Sustainable Wastewater Treatment Using Membrane Technology. Singapore: Springer Nature; 2023. DOI: 10.1007/978-981-19-9176-9_2

[20] Sacco NA, Zoppas FM, Devard A, del González Muñoz MP, García G, Marchesini FA. Recent advances in microplastics removal from water with special attention given to photocatalytic degradation: Review of scientific research. Microplastics. 2023;**2**(3):278-303. DOI: 10.3390/microplastics2030023

[21] Pellegrino J. Filtration and ultrafiltration equipment and techniques. Separation and Purification Methods. 2000;**29**(1):91-118. DOI: 10.1081/SPM-100100004

[22] Giacobbo A, Pasqualotto IF, de Machado Filho RCC, Minhalma M, Bernardes AM, de Pinho MN. Ultrafiltration and nanofiltration for the removal of pharmaceutically active compounds from water: The effect of operating pressure on electrostatic solute—Membrane interactions. Membranes (Basel). 2023;**13**(8):743. DOI: 10.3390/membranes13080743

[23] Othman NH et al. A review on the use of membrane technology systems in developing countries. Membranes (Basel). 2022;**12**(1):30. DOI: 10.3390/membranes12010030

[24] El Batouti M, Al-Harby NF, Elewa MM. A review on promising membrane technology approaches for heavy metal removal from water and wastewater to solve water crisis. Water (Switzerland). 2021;**13**(22):3241. DOI: 10.3390/w13223241

[25] Ma B, Yu W, Liu H, Qu J. Effect of low dosage of coagulant on the ultrafiltration membrane performance in feedwater treatment. Water Research. 2014;**51**:277-283. DOI: 10.1016/j.watres.2013.10.069

[26] Shobana KH, Ramya R, Mohan D. Development of antifouling ultrafiltration membranes by surface modification. International Journal of Nuclear Desalination. 2010;**4**(1):58-65. DOI: 10.1504/IJND.2010.033767

[27] Zheng W et al. Research on the factors influencing nanofiltration membrane fouling and the prediction of membrane fouling. Journal of Water Process Engineering. 2024;**59**:104876. DOI: 10.1016/j.jwpe.2024.104876

[28] Wang E et al. A selective separation mechanism for mono/divalent cations and properties of a hollow-fiber composite nanofiltration membrane having a positively charged surface.

Membranes (Basel). 2024;**14**(1):1. DOI: 10.3390/membranes14010001

[29] Alhussaini MA, Souza-Chaves BM, Felix V, Achilli A. Comparative analysis of reverse osmosis and nanofiltration for the removal of dissolved contaminants in water reuse applications. Desalination. 2024;**586**:117822. DOI: 10.1016/j.desal.2024.117822

[30] Cassano A, Rastogi NK, Basile A. Membrane Technologies for Water Treatment and Reuse in the Food and Beverage Industries. Elsevier Ltd; 2015. pp. 551-580. DOI: 10.1016/B978-1-78242-121-4.00018-6

[31] Tin MMM et al. Membrane fouling, chemical cleaning and separation performance assessment of a chlorine-resistant nanofiltration membrane for water recycling applications. Separation and Purification Technology. 2017;**189**:170-175. DOI: 10.1016/j.seppur.2017.07.080

[32] Jafari M et al. Cost of fouling in full-scale reverse osmosis and nanofiltration installations in the Netherlands. Desalination. 2021;**500**:114865. DOI: 10.1016/j.desal.2020.114865

[33] Wang XM, Li B, Zhang T, Li XY. Performance of nanofiltration membrane in rejecting trace organic compounds: Experiment and model prediction. Desalination. 2015;**370**:7-16. DOI: 10.1016/j.desal.2015.05.010

[34] Qasim M, Badrelzaman M, Darwish NN, Darwish NA, Hilal N. Reverse osmosis desalination: A state-of-the-art review. Desalination. 2019;**459**:59-104. DOI: 10.1016/j.desal.2019.02.008

[35] Wimalawansa SJ. Purification of contaminated water with reverse osmosis: Effective solution of providing clean water for human needs in developing countries international journal of emerging technology and advanced engineering purification of contaminated water with Reve. Certif. J. 2008;**9001**(12):75-89. [Online]. Available from: https://www.researchgate.net/publication/284804889

[36] Rodríguez-Calvo A, Silva-Castro GA, Osorio F, González-López J, Calvo C. Reverse osmosis seawater desalination: Current status of membrane systems. Desalination and Water Treatment. 2015;**56**(4):849-861. DOI: 10.1080/19443994.2014.942378

[37] Qurie M et al. Inland treatment of the brine generated from reverse osmosis advanced membrane wastewater treatment plant using epuvalisation system. International Journal of Molecular Sciences. 2013;**14**(7):13808-13825. DOI: 10.3390/ijms140713808

[38] Ahmed MA, Amin S, Mohamed AA. Fouling in reverse osmosis membranes: Monitoring, characterization, mitigation strategies and future directions. Heliyon. 2023;**9**(4):e14908. DOI: 10.1016/j.heliyon.2023.e14908

[39] Ladewig B, Al-Shaeli MNZ. Fundamentals of Membrane Bioreactors. Singapore: Springer; 2017. pp. 13-38. DOI: 10.1007/978-981-10-2014-8

[40] Chang X, Zhang C, He Y, Dong X, Jin W, Xu N. A comparative study of the performance of symmetric and asymmetric mixed-conducting membranes. Chinese Journal of Chemical Engineering. 2009;**17**(4):562-570. DOI: 10.1016/S1004-9541(08)60245-1

[41] Khorsand-Ghayeni M, Barzin J, Zandi M, Kowsari M. Fabrication of asymmetric and symmetric membranes based on PES/PEG/DMAc. Polymer Bulletin. 2017;**74**(6):2081-2097. DOI: 10.1007/s00289-016-1823-z

[42] Buonomenna MG. Mining critical metals from seawater by subnanostructured membranes: Is it viable? Symmetry (Basel). 2022;**14**(4):681. DOI: 10.3390/sym14040681

[43] Bachler S, Ort M, Krämer SD, Dittrich PS. Permeation studies across symmetric and asymmetric membranes in microdroplet arrays. Analytical Chemistry. 2021;**93**(12):5137-5144. DOI: 10.1021/acs.analchem.0c04939

[44] Tan XM, Rodrigue D. A review on porous polymeric membrane preparation. Part I: Production techniques with polysulfone and poly (vinylidene fluoride). Polymers. 2019;**11**(7):1160. DOI: 10.3390/polym11071160

[45] Wang S, Saffar A, Ajji A, Wu H, Guo SY. Fabrication of microporous membranes from melt extruded polypropylene precursor films via stretching: Effect of annealing. Chinese Journal of Polymer Science (English Edition). 2015;**33**(7):1028-1037. DOI: 10.1007/s10118-015-1643-x

[46] Dianat G, Movsesian N, Gupta M. Vapor deposition of functional porous polymer membranes. ACS Applied Polymer Materials. 2020;**2**(2):98-104. DOI: 10.1021/acsapm.9b01177

[47] Tan ECD, Sow S, Lock M, Bohr SJ, Barbe S, Wang F. State-of-the-Art Review of Porous Polymer Membrane Formation Characterization—How Numerical and Experimental Approaches Dovetail to Drive Innovation. Switzerland: Frontiers in Sustainability; 2023

[48] Kumar A, Chang DW. Optimized polymeric membranes for water treatment: Fabrication, morphology, and performance. Polymers (Basel). 2024;**16**(2):271. DOI: 10.3390/polym16020271

[49] Strathmann H, Kock K, Amar P, Baker RW. The formation mechanism of asymmetric membranes. Desalination. 1975;**16**(2):179-203. DOI: 10.1016/S0011-9164(00)82092-5

[50] Chen X, Yang Y, Li Y, Bao M, Zhang D, Wang Z. Integrated asymmetric superwetting Janus membrane for the efficient separation of various surfactant-stabilized oil-water emulsions. Environmental Science Nano. 2021;**8**(8):2235-2248. DOI: 10.1039/d1en00335f

[51] Jiang LY. In: Drioli E, Giorno L, editors. Asymmetric Membrane BT - Encyclopedia of Membranes. Berlin, Heidelberg: Springer Berlin Heidelberg; 2016. pp. 126-127. DOI: 10.1007/978-3-662-44324-8_1762

[52] Kim IC, Yoon HG, Lee KH. Formation of integrally skinned asymmetric polyetherimide nanofiltration membranes by phase inversion process. Journal of Applied Polymer Science. 2002;**84**(6):1300-1307. DOI: 10.1002/app.10452

[53] Pinnau I, Freeman BD. Formation and modification of polymeric membranes: Overview. ACS Symposium Series. 1999;**744**:1-22. DOI: 10.1021/bk-2000-0744.ch001

[54] Polotsky AE, Polotskaya GA. Study on top layer structure of composite membranes. Journal of Membrane Science. 1998;**140**(1):97-102. DOI: 10.1016/S0376-7388(97)00269-X

[55] Farahbakhsh J, Vatanpour V, Khoshnam M, Zargar M. Recent advancements in the application of new monomers and membrane modification techniques for the fabrication of thin film composite membranes: A review. Reactive and Functional Polymers. 2021;**166**:105015. DOI: 10.1016/j.reactfunctpolym.2021.105015

[56] Souza VC, Quadri MGN. Organic-inorganic hybrid membranes in separation processes: A 10-year review. Brazilian Journal of Chemical Engineering. 2013;**30**(4):683-700. DOI: 10.1590/S0104-66322013000400001

[57] Fard AK et al. Inorganic membranes: Preparation and application for water treatment and desalination. Materials. 2018;**11**(1):74. DOI: 10.3390/ma11010074

[58] Cot L et al. Inorganic membranes and solid state sciences. Solid State Sciences. 2000;**2**(3):313-334. DOI: 10.1016/S1293-2558(00)00141-2

[59] Amin SK, Abdallah HAM, Roushdy MH, El-Sherbiny SA. An overview of production and development of ceramic membranes. International Journal of Applied Engineering Research. 2016;**11**(12):7708-7721

[60] Ehsani M, Kalugin D, Doan H, Lohi A, Abdelrasoul A. Bio-sourced and biodegradable membranes. Applied Sciences. 2022;**12**(24):1-36. DOI: 10.3390/app122412837

[61] Fajal S, Dutta S, Ghosh SK. Porous organic polymers (POPs) for environmental remediation. Materials Horizons. 2023;**10**(10):4083-4138. DOI: 10.1039/D3MH00672G

[62] Bandehali S, Sanaeepur H, Ebadi Amooghin A, Shirazian S, Ramakrishna S. Biodegradable polymers for membrane separation. Separation and Purification Technology. 2021;**269**:118731. DOI: 10.1016/j.seppur.2021.118731

[63] Raaijmakers MJT, Benes NE. Current trends in interfacial polymerization chemistry. Progress in Polymer Science. 2016;**63**:86-142. DOI: 10.1016/j.progpolymsci.2016.06.004

[64] Xie W, Geise GM, Freeman BD, Lee H-S, Byun G, McGrath JE. Polyamide interfacial composite membranes prepared from m-phenylene diamine, trimesoyl chloride and a new disulfonated diamine. Journal of Membrane Science. 2012;**403-404**:152-161. DOI: 10.1016/j.memsci.2012.02.038

[65] Lv X et al. Tannic acid reinforced interfacial polymerization fabrication of internally pressurized thin-film composite hollow fiber reverse osmosis membranes with high performance. Desalination. 2022;**538**:115926. DOI: 10.1016/j.desal.2022.115926

[66] Li X et al. Regulating the interfacial polymerization process toward high-performance polyamide thin-film composite reverse osmosis and nanofiltration membranes: A review. Journal of Membrane Science. 2021;**640**(July):119765. DOI: 10.1016/j.memsci.2021.119765

[67] Martínez-Izquierdo L, Malankowska M, Téllez C, Coronas J. Phase inversion method for the preparation of Pebax® 3533 thin film membranes for CO_2/N_2 separation. Journal of Environmental Chemical Engineering. 2021;**9**(4):105624. DOI: 10.1016/j.jece.2021.105624

[68] Alam J, Dass LA, Alhoshan MS, Ghasemi M, Mohammad AW. Development of polyaniline-modified polysulfone nanocomposite membrane. Applied Water Science. 2012;**2**(1):37-46. DOI: 10.1007/s13201-011-0021-2

[69] Alkandari SH, Castro-Dominguez B. Advanced and sustainable manufacturing methods of polymer-based membranes for gas separation: A review. Frontiers in Membrane Science and Technology. 2024;**3**:1-26. DOI: 10.3389/frmst.2024.1390599

[70] Grosso D. How to exploit the full potential of the dip-coating process to better control film formation. Journal of Materials Chemistry. 2011;**21**(43):17033-17038. DOI: 10.1039/c1jm12837j

[71] Butt MA. Thin-film coating methods: A successful marriage of high-quality and cost-effectiveness—A brief exploration. Coatings. 2022;**12**(8):1115. DOI: 10.3390/coatings12081115

[72] Sazali N, Syafiq Sharip M, Ibrahim H, Shahir Jamaludin A, Norharyati Wan Salleh W. Influence of dip-coating times towards oxygen separation performance. IOP Conference Series Materials Science and Engineering. 2020;**788**(1):11-16. DOI: 10.1088/1757-899X/788/1/012035

[73] Suleman MS, Lau KK, Yeong YF. Enhanced gas separation performance of PSF membrane after modification to PSF/PDMS composite membrane in CO2/CH4 separation. Journal of Applied Polymer Science. 2018;**135**(1):1-9. DOI: 10.1002/app.45650

[74] Li G, Knozowska K, Kujawa J, Tonkonogovas A, Stankevičius A, Kujawski W. Fabrication of polydimethysiloxane (Pdms) dense layer on polyetherimide (pei) hollow fiber support for the efficient CO_2/N_2 separation membranes. Polymers (Basel). 2021;**13**(5):1-15. DOI: 10.3390/polym13050756

[75] Tan PC, Ooi BS, Ahmad AL, Low SC. Formation of a defect-free polyimide/zeolitic imidazolate framework-8 composite membrane for gas separation: in-depth analysis of organic–inorganic compatibility. Journal of Chemical Technology and Biotechnology. 2019;**94**(9):2792-2804. DOI: 10.1002/jctb.5908

[76] Ceratti DR, Louis B, Paquez X, Faustini M, Grosso D. A new dip coating method to obtain large-surface coatings with a minimum of solution. Advanced Materials. 2015;**27**(34):4958-4962. DOI: 10.1002/adma.201502518

[77] Raje A, Koll J, Schneider ES, Georgopanos P. A novel organic solvent-free method for manufacturing polyethersulfone hollow fiber membranes using melt extrusion. Journal of Membrane Science. 2023;**683**:121837. DOI: 10.1016/j.memsci.2023.121837

[78] Xie J, Xu R, Lei C. Uniaxial stretching induced pore nucleation and growth in row-nucleated crystalline hard-elastic polypropylene film: The effect of activation volume and stretching work. Polymer (Guildf). 2018;**158**:10-17. DOI: 10.1016/j.polymer.2018.10.047

[79] Sadeghi F, Ajji A, Carreau PJ. Analysis of microporous membranes obtained from polypropylene films by stretching. Journal of Membrane Science. 2007;**292**(1):62-71. DOI: 10.1016/j.memsci.2007.01.023

[80] Islam MS, Ang BC, Andriyana A, Afifi AM. A review on fabrication of nanofibers via electrospinning and their applications. SN Applied Sciences. 2019;**1**(10):1-16. DOI: 10.1007/s42452-019-1288-4

[81] Altun E, Ahmed J, Onur Aydogdu M, Harker A, Edirisinghe M. The effect of solvent and pressure on polycaprolactone solutions for particle and fibre formation. European Polymer Journal. 2022;**173**:111300. DOI: 10.1016/j.eurpolymj.2022.111300

[82] Jiang J et al. Arced multi-nozzle electrospinning spinneret for high-throughput production of nanofibers. Micromachines. 2020;**11**(1):1-8. DOI: 10.3390/mi11010027

[83] Sharma GK, James NR. Progress in electrospun polymer composite Fibers for microwave absorption and electromagnetic interference shielding. ACS Applied Electronic Materials. 2021;**3**(11):4657-4680. DOI: 10.1021/acsaelm.1c00827

[84] Yan B, Zhang Y, Li Z, Zhou P, Mao Y. Electrospun nanofibrous membrane for biomedical application. SN Applied Sciences. 2022;**4**(6):172. DOI: 10.1007/s42452-022-05056-2

[85] Apel P. Track etching technique in membrane technology. Radiation Measurements. 2001;**34**:559-566. DOI: 10.1016/S1350-4487(01)00228-1

[86] Zdorovets MV, Kozlovskiy AL, Harrison FE, Tuleushev AZ. Induced ordering in polyethylene terephthalate films irradiated with Ar ions with an energy of 70 MeV. Surface and Coatings Technology. 2020;**386**:125490. DOI: 10.1016/j.surfcoat.2020.125490

[87] Ferain E, Legras R. Pore shape control in nanoporous particle track etched membrane. Nuclear Instruments and Methods in Physics Research Section B: Beam Interactions with Materials and Atoms. 2001;**174**(1-2, 122):116. DOI: 10.1016/S0168-583X(00)00455-9

[88] Kaya D, Keçeci K. Review—Track-etched Nanoporous polymer membranes as sensors: A review. Journal of the Electrochemical Society. 2020;**167**(3):037543. DOI: 10.1149/1945-7111/ab67a7

[89] Sa-Nguanruksa J, Rujiravanit R, Supaphol P, Tokura S. Porous polyethylene membranes by template-leaching technique: Preparation and characterization. Polymer Testing. 2004;**23**(1):91-99. DOI: 10.1016/S0142-9418(03)00066-7

[90] Chang HH, Yao LC, Lin DJ, Cheng LP. Preparation of microporous poly(VDF-co-HFP) membranes by template-leaching method. Separation and Purification Technology. 2010;**72**(2):156-166. DOI: 10.1016/j.seppur.2010.02.001

[91] Dommati H, Ray SS, Wang JC, Chen SS. A comprehensive review of recent developments in 3D printing technique for ceramic membrane fabrication for water purification. RSC Advances. 2019;**9**(29):16869-16883. DOI: 10.1039/c9ra00872a

[92] Low ZX, Chua YT, Ray BM, Mattia D, Metcalfe IS, Patterson DA. Perspective on 3D printing of separation membranes and comparison to related unconventional fabrication techniques. Journal of Membrane Science. 2017;**523**(2016):596-613. DOI: 10.1016/j.memsci.2016.10.006

[93] Tijing LD, Dizon JRC, Ibrahim I, Nisay ARN, Shon HK, Advincula RC. 3D printing for membrane separation, desalination and water treatment. Applied Materials Today. 2020;**18**:100486. DOI: 10.1016/j.apmt.2019.100486

[94] Aliyu UM, Rathilal S, Isa YM. Membrane desalination technologies in water treatment: A review. Water Practice Technology. 2018;**13**(4):738-752. DOI: 10.2166/wpt.2018.084

[95] Bhattacharjee C, Saxena VK, Dutta S. Fruit juice processing using membrane technology: A review. Innovative Food Science and Emerging Technologies. 2017;**43**:136-153. DOI: 10.1016/j.ifset.2017.08.002

[96] Ağtaş M, Yılmaz Ö, Dilaver M, Alp K, Koyuncu İ. Hot water recovery and reuse in textile sector with pilot scale ceramic ultrafiltration/nanofiltration membrane system. Journal of Cleaner Production.

2020;**256**:120359. DOI: 10.1016/j.jclepro.2020.120359

[97] Ahmad T et al. Unveiling the potential of membrane in climate change mitigation and environmental resilience in ecosystem. Environmental Research. 2024;**245**:117960. DOI: 10.1016/j.envres.2023.117960

[98] Gohil PP, Desai H, Kumar A, Kumar R. Current status and advancement in thermal and membrane-based hybrid seawater desalination technologies. Water (Switzerland). 2023;**15**(12):2274. DOI: 10.3390/w15122274

[99] Esmaeilion F. Hybrid renewable energy systems for desalination. Applied Water Science. 2020;**10**(3):84. DOI: 10.1007/s13201-020-1168-5

[100] Fellows CM, Al-Hamzah A. Thermal Desalination: Current Challenges. Amsterdam: Elsevier; 2015. p. 2015. DOI: 10.1016/B978-0-444-63228-9.00023-1

[101] Wu J et al. Polyamide reverse osmosis membrane compaction and relaxation: Mechanisms and implications for desalination performance. Journal of Membrane Science. 2024;**706**:122893. DOI: 10.1016/j.memsci.2024.122893

[102] Piash KPS, Sanyal O. Design strategies for forward osmosis membrane substrates with Low structural parameters—A review. Membranes (Basel). 2023;**13**(1):73. DOI: 10.3390/membranes13010073

[103] Lachish U. Optimizing the efficiency of reverse osmosis seawater desalination. 2002:1-17. DOI: 10.13140/RG.2.2.12099.20003. Available from: www.urila.tripod.com/ Seawater.htm [Accessed: December 15, 2010]

[104] Ayaz M, Namazi MA, ud Din MA, Ershath MIM, Mansour A, Aggoune E-HM. Sustainable seawater desalination: Current status, environmental implications and future expectations. Desalination. 2022;**540**:116022. DOI: 10.1016/j.desal.2022.116022

[105] Goh PS, Lau WJ, Othman MHD, Ismail AF. Membrane fouling in desalination and its mitigation strategies. Desalination. 2018;**425**:130-155. DOI: 10.1016/j.desal.2017.10.018

[106] Shaban M, El Sayed H, Abdel-Hamid H, Morsy A, Kandil S. Anti-biofouling of 2-acrylamido-2-methylpropane sulfonic acid grafted cellulose acetate membranes used for water desalination. Chemical Engineering and Processing Process Intensification. 2020;**149**:107857

[107] Abd-El-Khalek DE, Abd-El-Nabey BA, Morsy A, Ebrahim S, Ramadan SR. Improvement of performance and antifouling properties of reverse osmosis membranes using green additive. Desalination and Water Treatment. 2019;**142**:65-71

[108] Abd-El-Khalek DE, Ramadan SR, Morsy A, Abd-El-Nabey BA. Investigation of antifouling properties of modified reverse osmosis membrane containing natural extract. Journal of Materials and Environmental Science. 2021;**12**(6):812-822

[109] Fadl E, Noumir H, Morsy A, Kandil S, Ebrahim S, Khalil MM. Development of cellulose acetate membrane performance by carboxylate multiwalled carbon nanotubes. Advances in Natural Sciences: Nanoscience and Nanotechnology. 2022;**13**(1):015006

[110] Panigrahy SK, Nandha A, Chaturvedi M, Mishra PK. Novel nanocomposites with advanced materials

and their role in waste water treatment. Next Sustainability. 2024;**4**:100042. DOI: 10.1016/j.nxsust.2024.100042

[111] Bandehali S et al. Planning of smart gating membranes for water treatment. Chemosphere. 2021;**283**:131207. DOI: 10.1016/j.chemosphere.2021.131207

[112] Zhang J, Soliman A, El-marghany A, Morsy A, Mohamed A, Shehata N, et al. Utilization of MOF enhanced hydrophilic nanocomposite reverse osmosis membrane for desalination with antifouling capabilities. Polymer Engineering and Science. 2024;**64**(9):4430-4441. DOI: 10.1002/pen.26858

[113] Morsy A, Mahmoud AS, Soliman A, Ibrahim H, Fadl E. Improved anti-biofouling resistances using novel nanocelluloses/cellulose acetate extracted from rice straw based membranes for water desalination. Scientific Reports. 2022;**12**(1):4386

[114] Rabiee N et al. Green and sustainable membranes: A review. Environmental Research. 2023;**231**(P2):116133. DOI: 10.1016/j.envres.2023.116133

Chapter 5

Adsorbents for Water Desalination

Vishwakarma Ravikumar Ramlal and Savan K. Raj

Abstract

The necessity for freshwater is growing as the global population continues to expand. One of the practices the scientific community has proposed to address the present global freshwater crisis is water desalination. This process promotes the production of fresh water from salty water. Due to the significance of high salt removal efficiency, cheap cost, minimal environmental effect, and comparatively low energy requirement, adsorption is considered a potential method for desalination. Predominantly, adsorption techniques do not use chemicals. Among the frequently studied adsorbents for desalination are activated carbons, zeolites, carbon nanomaterials, graphene, and metal or covalent organic framework materials. These materials exhibit various capabilities in terms of adsorption rate, adsorption capacity, stability, and recyclability. Carbon nanotubes (CNTs) and graphene, two next-generation materials that show numerous functions with increased water transport capabilities, play a significant role and have been considered very appealing enhancers to the desalination process. However, most functional materials have drawbacks, including the need for specialized synthesis methods, agglomeration, leaching, and issues related to the environment and human health. This chapter will focus on current trends in adsorbent material development and evaluate the most recent materials with their properties, which might help with adsorbent design from an engineering application standpoint.

Keywords: adsorbents, porous materials, adsorption desalination, graphene, zeolite, MOFs, and COFs

1. Introduction

Freshwater resources and energy supplies are under increasing stress due to modern society's fast expansion and growing population [1, 2]. Urbanization, economic growth, and rising living standards have made the water shortage a worldwide issue [3]. The demand for freshwater will have increased by 40% by 2050 [4–6]. Most of the world is covered with seawater, making extracting fresh water from it practical. As a result, desalination technology emerges as a viable solution to the issue of water shortage [7, 8]. Nowadays, three primary varieties of desalination techniques are commonly utilized globally: (1) thermal desalination, which includes membrane desalination techniques, like forward osmosis (FO) and reverse osmosis (RO); (2) chemical desalination, which comprises gas hydrate and ion exchange desalination; and (3) non-thermal desalination, that involves multi-effect distillation (MED) and multi-stage flash evaporation [9–11]. Reverse osmosis and multi-effect distillation are the two most widely used desalination methods available today [12, 13]. It is interesting to note that various desalination techniques are applied in various parts of the world and that the ratio of applications varies as well (**Figure 1**). However, there are still significant problems with traditional desalination methods, including: (1) increased maintenance costs; (2) corrosion and blockage of membrane mass exchangers; and (3) high-grade fossil power consumption and pollution emission. As such, classic desalination techniques are not up to the current demands of sustainable energy and environmental development.

A thin layer of molecules, ions, or atoms from a material (gas or liquid) sticks to a surface and forms adsorption, a surface phenomenon. It is used for contamination removal, separation, and purification and involves either chemical or physical interactions. The necessity of creating a unique desalination methodology that uses little power and produces negligible pollution is growing considering the necessity of sustainable human development. As seen in **Figure 2**, adsorption desalination (AD) is a recent method that achieves desalination by utilizing certain solid materials with porous qualities to both adsorb and desorb water vapor [14, 15]. Because of its great selectivity, effectiveness at low concentrations, and adaptability in eliminating a variety of pollutants, adsorption is a technique that shows promise. Because many adsorbents can be recycled and regenerated, it is both cost-effective and ecologically benign, making them suitable for a wide range of applications.

A variety of porous nanomaterials can be utilized in the adsorption desalination systems (ADS), like silica gel, zeolite, and MOFs, and all consist of excellent sorption capacity. These porous materials consist of unique characteristics, including high pore volume and specific area as well as fluid permeability, excellent adsorption capacity, quick adsorption kinetics, and strong selectivity [16–19]. Since adsorption is associated with the interconnection of the foreign molecules with the surface of the adsorbent, a large surface area is often regarded as a crucial attribute for the enhancement of the sorption capability. However, the guest molecules must have access to

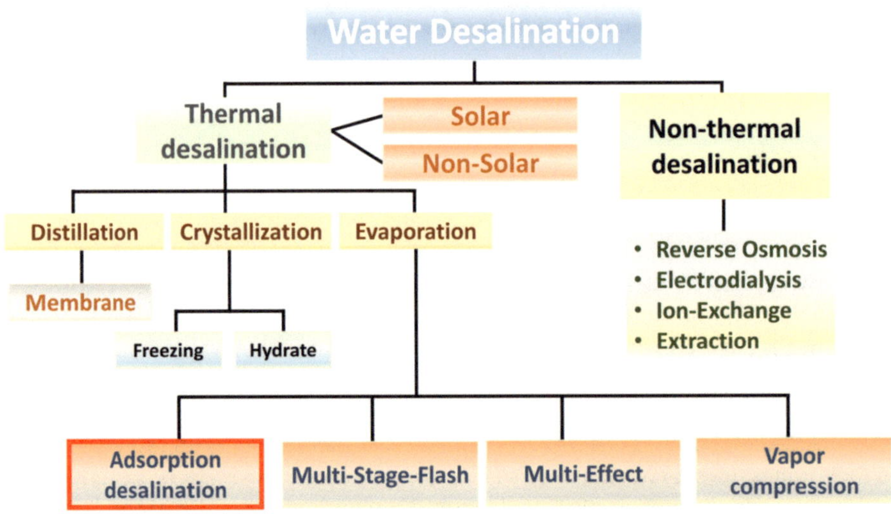

Figure 1.
Categories of presently available desalination technology [14].

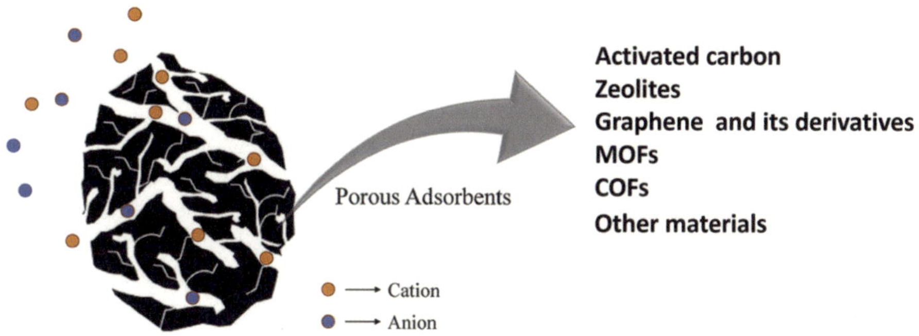

Figure 2.
Adsorption and desorption process for porous materials [15].

the porous material's surface, particularly its interior surface. Pore sizes determine the material's surface area, i.e., as the average pore size decreases the surface area increases [20, 21]. The average pore size of the material must be greater than the kinetic diameter of the foreign molecules to allow rapid adsorption kinetics. If the material has secondary pores, a hierarchical pore network is required to provide better guest accessibility to the host surface [22, 23]. Given its potential as a solution to the water scarcity issue, this chapter offers an explanatory overview of the state-of-the-art adsorption process, including the most recent porous contender nanomaterials used as adsorbents [24]. This assessment can serve as a roadmap for future studies aimed at creating highly efficient adsorption desalination technology for real-world applications.

2. Overview of the emerging and well-known adsorbents

2.1 Activated carbon (AC)

Activated carbon is considered as a category of facile carbon-based materials, that can be attained via different economically affordable raw materials like wood straws, glucose, nut shells, coal, scrap plastic, papermaking waste, and sewage waste [25–27]. The porosity characteristics of AC can be tuned according to the activation methods and precursors utilized. Because of their excellent porosity, vast surface area, different surface functional groups, etc., AC with hierarchical porous construction could be highly advantageous for adsorption [28, 29]. The structure and chemical characteristics of AC must be constructed for the enhancement of its adsorption capability. This can be attained by a variety of post-synthetic modification methods, including oxidation, that can be attained via chemical, air, or electrochemical oxidation [30–37]. The most popular technique for oxidizing AC is acid treatment with nitric and sulfuric acids [38, 39], which can provide the altered AC surface oxygenated groups. These different groups interact with metal ions for the formation of metal complexes, which increase the capability of AC to absorb metal ions [40–42]. It was observed that the steadiness of the metal complexes that formed correlated with the affinity of metal ions to the AC (acid-treated) [43]. Because of dipole–dipole interactions, hydrogen bonds, and covalent bonds, grafted AC containing nitrogen (N) functionalities, deliver a comparatively high adsorption capability for negatively charged ions

[44, 45]. Organic species from water, like 4-chlorophenol and 2,4 dichlorophenol (2,4-DCP), may be effectively eliminated by using base-treated AC [46]. For instance, there was a 22.8% improvement in the remediation of 2,4-DCP on AC functionalized with ammonia [47]. The active surface area and pore volumes are often reduced by both basic and acid treatments for AC, which may harm adsorption. For instance, even though the NaOH-treated AC had more oxygen groups on it, the sorption ability of AC for chromium ions was not increased [48]. This might be lessened by tuning the analysis parameters. For instance, it has been reported that following thermal ammonia treatment, AC's porosity and pore structure were maintained, which helped boost the adsorption capacity for perchlorate (ClO_4) [49]. In Addition, microwave irradiation treatment has been used for AC, offering rapid and selective heating that cuts down on treatment time and energy [50–52]. Furthermore, in other reports treatment using plasma and ozone was employed [53–56]. However, it should be considered that AC sorption is primarily utilized as electrodes in the capacitive deionization (CDI) method for deionizing water sources. It was also frequently utilized to remediate heavy metals and organic pollutants, and there was only a limited evaluation of the adsorption capability of AC for ionic salts (like sodium and chloride) [57, 58].

2.2 Zeolites

Natural zeolites are employed as desalination adsorbents [59, 60]. Zeolites have a well-organized three-dimensional (3D) architecture with corals and channels within. Their general chemical formula is $M_{x/n}$ $[Al_xSi_yO_{2\,(x+y)}].pH_2O$, where M is (Na, K, Li) and/or (Ca, Mg, Ba, Sr), n is the cation charge, and y/x = 1–6 and p/x = 1–4. Because silicon and aluminum contain valences that make them inherently negatively charged, zeolites may exchange ions with a wide range of positively charged organic compounds and cations. Furthermore, the high porosity of zeolite (usually 30–40% with pore diameters of 0.3–0.8 nm) can facilitate the high adsorption kinetically [61, 62]. Since zeolite water may be considered as a pathway for the exchange of ions between exchangeable cations and ionic salts, it can also accelerate the adsorption kinetics. Different types of zeolites exist in nature (like phillipsite, chabazite, stilbite, analcime, clinoptilolite, and mordenite) with different adsorption properties. The crystalline structure and constituent present within have a direct impact on their adsorption capability and kinetics. For instance, mordenite type zeolite, a common natural zeolite, has an excellent ability to adsorb Ca^{2+} and Mg^{2+} due to ion transportation between sodium ion and the two cations, and may efficiently decrease the concentration of Na^+ in the region i.e. around 73% [63, 64]. However, the negative charge of zeolite will reduce the adsorption capacity of anions like Cl^- and SO_4^{2-} by 20% and 30%, respectively [65]. With an exchange capacity of 11 mg g^{-1}, another naturally occurring zeolite bigadic clinoptilolite, can also remove Ca^{2+} from aqueous media [63]. Seawater desalination is another use for natural zeolites. In a study, Wajima et al. [65] extracted NaCl from seawater by utilizing a modified zeolite from Fukushima, Japan. Before adsorption, the zeolites were functionalized with silver nitrate and heated at 60°C for 12 h. This study resulted in an 81.73% declination in NaCl under batch conditions. In a literature report, Wibowo et al. [66] used functionalized zeolites from Indonesia (clinoptilolite) to decrease the salinity in seawater. The adsorption ability of the zeolite was increased to 51.43 mg g^{-1} in the lab-scale pilot method by thermally activated at 225°C for 3 h with no usage of any chemical reagents. Zeolites have a sorption ability of 2–20 mg g^{-1}, which is often more than that of AC [63]. Therefore, natural zeolites may be an affordable option for saltwater

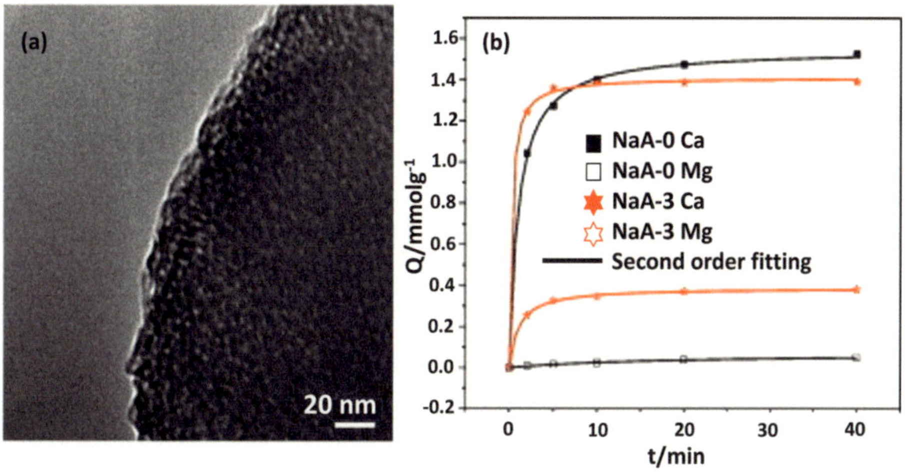

Figure 3.
(a) TEM image of the zeolite, and (b) pseudo-second-order fitting lines and competitive exchange data of Ca^{2+}/ Mg^{2+} [75].

desalination sorbent materials. Zeolites are useful in water treatment for ionic salts and heavy metal ions as well as inorganic anions (e.g., NO^{3-}, SO_4^{2-}, PO_4^{3-}, F^-, ClO_4^-, CN^-) [67–70] and organic compounds, like dyes [71–74]. Hydrothermally produced zeolites, like (Linde type A) LTA, can also be employed as adsorbents (**Figure 3**). The physicochemical properties of these materials can be adjusted for the enhancement of their remediation ability by altering the synthesis conditions. For example, mesoporous LTA (mesopores with pore size 3 nm) was created to decline the diffusion resistance. It illustrates an improved Mg^{2+} adsorption rate because of the increased ion diffusion rate in its porous network, which was greater than the pristine microporous LTA by about 17.5 times [75].

2.3 Graphene and graphene-based materials

Graphene and its related materials are well-known for water treatment applications. When graphene is pristine, it consists of a singular sheet of carbon atoms situated in an aromatic sp^2 hybridization. With a theoretical surface area value of around 2630 m^2g^{-1}, graphene is considered a popular example of a high-surface material. The reason is that each atom in a graphene sheet consisting of a singular layer is open to the environment on both ends. Graphene is a desirable material for adsorption technology because of its large surface area, rich in oxygen content, adjustable surface chemistry, and scalable manufacture [76–82]. **Figure 4** shows the composition and **Table 1** reflects the key characteristics of graphene-based materials (like graphene oxide (GO)) employed in water treatment applications. The oxygen-containing groups in graphene are responsible for the remediation of metal ions. Because GO comprises oxygen species that can interact with metal species, it is frequently chosen over pure graphene for metal ion remediation [83]. The remediation capacity of graphene can be progressively impacted by ionic strength and pH in addition to oxygen-containing groups.

The electric double layer (EDL) of hydrated chemical functionalities can change depending on the ionic strength, which can affect how metal ions attach to GO sheets. In general, reducing pH values causes the sorption capacity to decline. An inclination

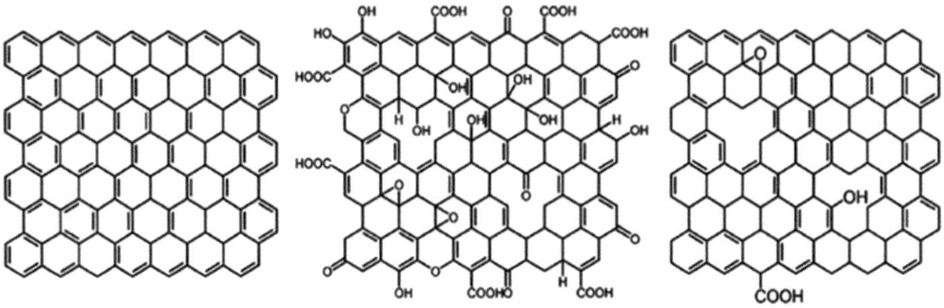

Figure 4.
An overview of structural properties of graphene and its derivatives for adsorption and desalination [83].

Properties	Graphene	Graphene oxide	Reduced graphene oxide
Synthesis	• Chemical vapor deposition • Thermal decomposition of SiC • Graphite exfoliation	Oxidation and exfoliation of graphite	Reduction of graphene oxide
C:O ratio	No oxygen	2–4	8–246
Young's modulus (TPa)	1	0.2	0.25
Electron mobility ($cm^2V^{-1}s^{-1}$)	10,000–50,000	insulator	0.05–200
Production cost	High	Low	Low

Table 1.
Properties of graphene and its derivatives [83].

in adsorption capacity results from the positive charge of metal ions interacting favorably with GO's negatively charged surface. Graphen-based materials have also been used to remediate anionic species from aqueous media, like Cl^-, PO_4^-, ClO_4^-, and F^-. The process of anionic species sorption was linked to anion-π interactions, in contrast to the immobilization of cationic metal species. Mishra et al. [84] used hydrogen-induced exfoliation of GO to produce and functionalize graphene sheets. Using the functionalized adsorbent, monovalent ions (sodium) and multivalent ions (arsenic (III and V)) were remediated from an aqueous media at the same time. The highest sorption capacities for sodium and arsenic were 122, and 142 mg g^{-1}, respectively, according to Langmuir isotherm. GO has a strong adsorption capacity to remove divalent ions (like calcium) from drinking water [85]. Overall, GO helps to remove monovalent and divalent ions up to 99% of the time. According to experimental studies, GO has a removal efficiency of 212 mg g^{-1}. In comparison to other adsorbents for example AC and activated alumina, these values are quite high. Given its comparable chemistry, it is anticipated that GO will similarly be efficient in eliminating magnesium ions from water. The free-standing GO membranes have been used excessively for desalination purposes (**Figure 5(a)**) [86]. In another report, Chu et al., fabricated a new β-cyclodextrin (β-CD) modified graphene oxide (CDGO) membrane for the effective adsorption of bisphenol A (BPA), with a high flux and sorption capacity [87]. Three CDGO membranes with variations in the concentration of CDGO nanosheets i.e. 5 mg, 10 mg, and 20 mg were made and labeled as M1, M2, and M3 (**Figure 5(b)**).

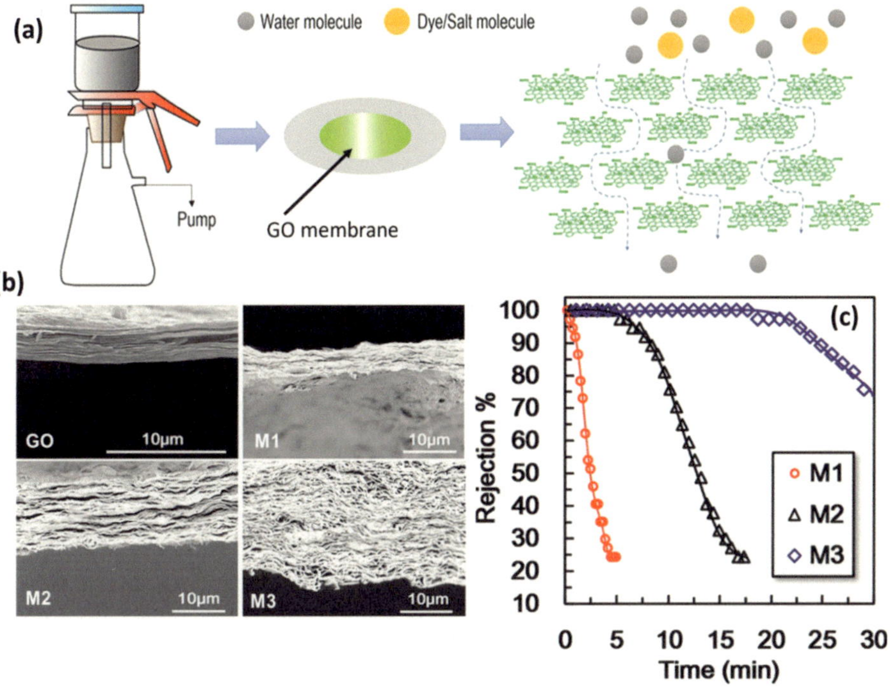

Figure 5.
(a) Process of making the GO self-standing membrane, (b) SEM images of the self-standing GO membrane, and (c) BPA removal performances of M1, M2, and M3 membranes in the 0.01 mM BPA [85, 86].

These membranes have fluxes one or two orders of magnitude greater than the reverse osmosis (RO) and nanofiltration (NF) membranes earlier utilized for BPA remediation, with a BPA rejection effectiveness of almost 100% (**Figure 5(c)**).

Since their discovery by Iijima [88], carbon nanotubes (CNTs) have been the attractive subject of much research for a variety of uses, including desalination. CNTs are one-dimensional (1D) cylinders made of micro or nanoscale graphene sheets. Their high porosity, large surface areas, easy functionalization, and hollow/layered architectures make them a perfect fit for desalination and water purification applications [89–91]. For instance, in a study, CNT nanocomposite sheets were created via the chemical vapor deposition (CVD) technique for sodium ion remediation. Even so, because of the low loading of CNTs on the substrate, the adsorption capacity was very restricted up to 22.09 mg g^{-1} [92]. Forming CNT sheets aligned in the vertical direction with functionalized open ends that permit liquid to get through is an additional method [93, 94]. Because of the decreased friction on CNT walls, there may be a significant water permeance. More significantly, because ionic salts are absorbed rather than remediation, a low mass transfer resistance can be shown, implying a shallow operating pressure and minimal power usage. Yang et al. [95] provided evidence of CNT sheets' ability to remove salt from aqueous conditions. They demonstrated that plasma treatment of CNTs produced a salt removal capacity that was two orders of magnitude higher than that of AC, surpassing 400% by mass. It is important to note that producing CNT sheets with high surface areas is extremely difficult, even if doing so is essential to moving the technology closer to commercialization [96]. **Figure 6** illustrates the methods to increase the sorption capability of CNTs.
In another study, Cao et al. used a ball-milling method to synthesize nickel

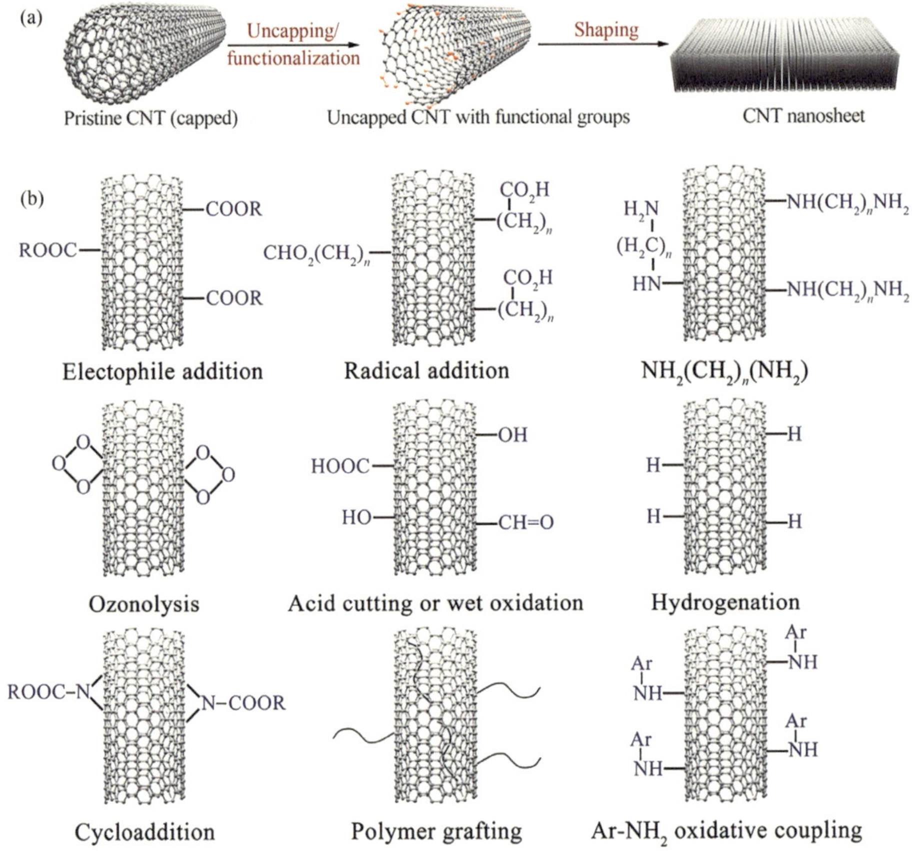

Figure 6.
Methods to increase the sorption ability of CNTs: (a) popular method for using CNTs for the adsorption process; and (b) various methods for surface modification of CNTs [96].

hexacyanoferrate and carbon nanotube composites (NiHCF@CNTs) [97]. They demonstrate an asymmetric flow electrode capacitive deionization system (AFCDI) employing NiHCF@CNTs and AC as the flow cathode and anode, respectively. These results indicated that the AFCDI system outperformed the Flow electrode capacitive deionization (FCDI) system in terms of average salt remediation rate and power consumption. This was because the combination of the high specific capacitance of NiHCF and the superior conductivity of CNTs improved charge transfer.

2.4 Metal: organic frameworks (MOFs)

Metal–organic framework nanomaterials (MOFs) are formed by coordination-bonded organic ligands and metal clusters. They are well-known for having uniform pore sizes, high porosity, tunable physiochemical properties, large specific surface area (1000–10,000 m^2g^{-1}), and the capability to be functionalized easily [98–100]. With extraordinarily high adsorption efficiencies, MOFs are extremely attractive next-generation adsorbents due to their beneficial properties [101–103]. A variety of anionic substrates, including ClO_4^-, MnO_4^{2-}, CrO_4^-, and ReO_4^- [104–107], can be efficiently adsorbed by MOFs possessing cation charge capability, such as SLUG-21, SLUG-22, SLUG-35, and 1-ClO_4. Several Zn-based MOFs were reported for the remediation of AsO_4^{3-} and $Cr_2O_7^{2-}$, and some Chromium-based MOFs demonstrated high sorption capability for some anionic adsorbates, including F^-, AsO_3^{4-}, SeO_4^{2-}, PO_4^{3-}, SO_4^{2-}, and $PtCl_6^{2-}$ [108, 109]. Anion exchange adsorption can also be accomplished using 3D cationic MOFs [110–112]. For their use in desalination, the stability of MOFs in aqueous conditions presents challenges. Numerous MOFs with strong adsorption properties and comparatively high water stability were fabricated during these periods, as shown in **Figure 7**.

Three main approaches can be used to enhance the sorption capacity: (i) synthesize MOFs with lengthy organic molecules to increase the size of the pore and aid in the passing of substances within MOFs; (ii) design flawed MOFs to increase their pore size and/or add more sorption cavities [113–115]; and (iii) functionalize MOFs. In particular, the surface modification of MOFs can enhance their sorption/exchange and regeneration properties. MOFs can be functionalized in two ways: (i) by grafting charge-balancing anionic species (like polyelectrolytes) to their metal sites. This can be achieved by modifying the synthesis anion stripping process and adding charged secondary building units during and after synthesis [116], and (ii) by introducing counterions to ligands of MOFs through self-assembly. Additional pendant functional groups, such as —NH_2, —OH, and —SH, can increase the number of sorption sites and improve the selectivity of MOFs [117, 118]. Although MOFs have been utilized in CDI [119, 120], they cannot be directly utilized as electrode materials because of their minimal electric conductivity and chemical stability. To achieve conductivity, the controlled porosity coming from the structure of the parent MOFs is utilized [121], carbon nanomaterials were produced employing MOFs (as templates) for CDI applications. In a study, a 3D porous MOF-199@PVP/MWCNTs (CMPC) was synthesized by Li et al. to improve U(VI) electrosorption in water [122]. Multi-walled carbon nanotubes (MWCNTs) interpenetrating one another created a localized conductive

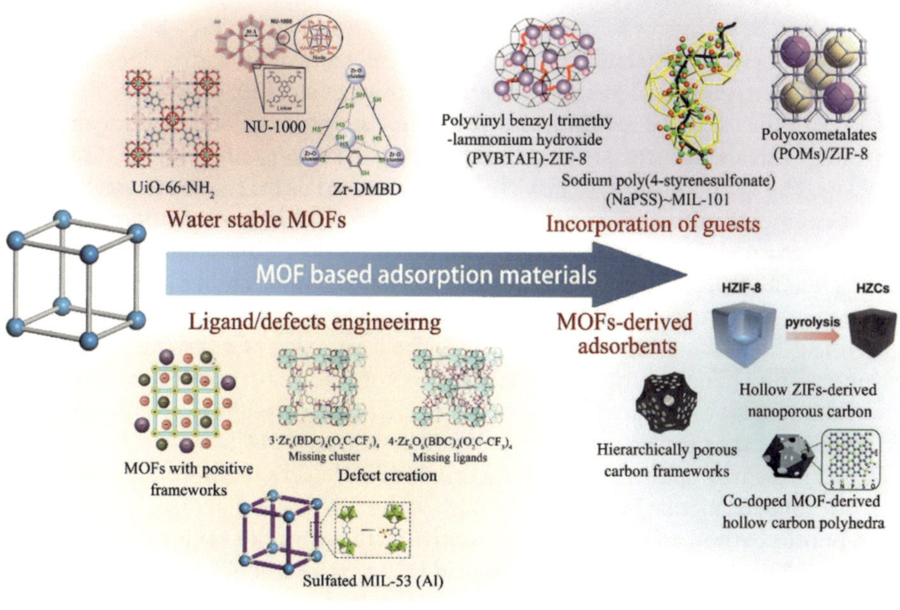

Figure 7.
Different types of MOFs for the water treatment application [96].

network that improved their characteristics even further. With an impressive removal ratio of 95.2% and an uptake electrosorption ability of 410.3 mg g^{-1} for UO_2^{2+} aqueous media (1.2 V), the CMPC electrode was thus proved to be effective. In another work, Peng et al. fabricated NiFe-metal organic framework@NiFe-layered double hydroxides/carbon fiber (MOF@LDH/CC) electrodes for effective electrosorption [123]. In order to prevent the aggregation of NiFe-LDH and enhance the bond between the CC and NiFe-LDH, the NiFe-MOF acts as a catalyst to facilitate the growth of LDH. For preventing aggregation of NiFe-LDH and strengthening a bond between the CC and NiFe-LDH, the NiFe-MOF serves as a seed to promote the growth of LDH. The improved structures performed well and attained an excellent initial salt remediation capacity of 0.033 mg cm^{-2}.

2.5 Covalent organic framework (COFs)

Covalent bonds between light elements like carbon (C), nitrogen (N), hydrogen (H), oxygen (O), and boron (B) result in the formation of crystalline framework materials, or COFs [124]. Their intrinsic properties, like excellent surface area, uniform tunable aperture diameters, and great chemical stability, make them suitable nanomaterials for adsorption applications. The removal of salts, heavy metals, and dyes has all been done using COFs [125–129]. Lewis acid–base interactions, hydrogen bonds, π–π stacking, and electrostatic attractions are the primary mechanisms behind dye molecule adsorption on COFs. For instance, at tiny concentrations of 32×10^{-6} mol L^{-1}, highly water-stable polycationic COFs can remediate many organic dyes, dominantly through electrostatic interaction between the anionic groups of the dyes and the bipyridinium cations of the COF [130]. High sorption capacities for dyes like auramine O and rhodamine B were also demonstrated by the collaboration of MOF-5 and melamine-based COFs [131]. Additionally, COFs have shown the ability to adsorb heavy metals, including Pb^{2+}, Hg^{2+}, U^{6+}, and Cr^{3+} [132–135]. Any prospective use of COFs requires their water stability, and many of them have weak stability in aqueous solutions, particularly those made of boroxines or boronate esters, that are readily countered by nucleophiles. Linkers that are relatively stable in water, such as imine, hydrazine, triazine, and azine, can be utilized to prepare COFs and increase their water stability [136–141]. To increase the compatibility of COFs in hostile environments, an intramolecular hydrogen bonding strategy can also be used [142, 143]. Enhancing the water stability of COFs can also be achieved by incorporating them with different materials that are water-stable. When GO was combined with COF, which is made up of 2,6-diaminoanthraquinone (DAAQ) and 1,3,5-triformylphloroglucinol (TFP), for instance, the resultant graphene-synergized COF (GS-COF) demonstrated better uranium and plutonium sorption capabilities than both GO and COF. For example, uranium 220.1 mg g^{-1} (GS-COF) compared to 92.5 mg g^{-1} (GO) and 105.0 mg g^{-1} (COF), The interaction of π systems between the contributing nanomaterials and various types of sheet structure, together with the chemical interaction produced throughout the intricate procedure, also enhanced the acid stability of the resulting complex in liquid conditions [144].

Present research has investigated the utilization of COFs in the elimination of salt from aqueous media, including brackish water [145]. For instance, a nitrogen-doped porous carbon anode and a redox-active COF with a salt removal efficiency of 22.8 mg g^{-1} have been carried out to a hybrid CDI as the cathode material [146]. Due to their high flexibility, COFs have the potential to be further explored and may offer unique insights into the advancement of materials for adsorption, even though their adsorption capability is currently limited when compared to other materials [147].

2.6 Other adsorbents

Different materials, including industrial wastes and by-products like fly ash, and waste slurry, were also illustrated for adsorption [148], in addition to the adsorbents covered above. For instance, fly ash from power plants is far more cost-effective and demonstrated a somewhat higher Na^+ elimination effectiveness (23%) than zeolite Y (21%) [149]. Agricultural wastes and by-products, such as fruit stones and/or shells, in their unprocessed or treated states, have also been studied as low-cost bio-adsorbents for treating wastewater and water [148]. One cost-effective aspect of desalination is the utilization of waste materials. However, their capabilities need to be enhanced before commercialization. Because of their high surface area, porous network, and strong hydrophilicity, 2D inorganic MXenes have gained popularity recently in various applications [149–155]. These have lately been illustrated as adsorbents for applications involving the removal of salts. For instance, MXene $Ti_3C_2T_x$ film was fabricated by Zhang et al., and remediation rates of Zn^{2+}, Pb^{2+}, phenol, and crystal violet by MXene composite membranes are above 90% [156]. The results indicate MXene's potential for desalination, which warrants more research and development. Agricultural wastes and byproducts have garnered a lot of interest lately as an adsorbent for the extraction of metal contaminants from aqueous medium. Bio-sorption is the term for using agricultural wastes and by-products in the bioremediation of pollutants like salt and heavy metals [157]. Hemicellulose, lignin, lipids, proteins, simple sugars, water, hydrocarbons, and starch are the common components of biological and agricultural wastes; each of these substances has a distinct functional category. These bio-materials could be employed in their original state or following chemical or physical alteration [158]. Moreover, silica-based nanomaterials have been broadly utilized to remediate ions from brackish and seawater [159]. To remove boron, an investigation was carried out by Amor et al. [160] in which they prepared homogeneous porous silica (OMS) type SBA-15 that was both pristine and treated using glucamine. In the experiments conducted, after three hours, the modified SBA-15 reached equilibrium and had a boron removal capability of around 45%, compared to 23% for the original one. Another study investigated for the remediation of boron using diol-functionalized silica by Tang et al. with excellent sorption capacities of 1.54 mmol g^{-1} [161] and N-methyl-D-glucamine based silica-supported adsorbent (Si-MG) by Xu et al., [162] which has the sorption capacity of 13.26 mg g^{-1}. Copper sulfate ($CuSO_4$), generally a salt hydrate substance, because of its reversible chemical interaction with water vapor, may be utilized in adsorption desalination. Ali et al. conducted a study to investigate the benefits of using copper sulfate salt hydrate as a new adsorption pair in a thermally driven adsorption system [163]. Also, functional materials are very much utilized in desalination processes that could either enhance the performance or lower the water production costs [164]. Functional materials are unquestionably crucial in many desalination systems. Nevertheless, these materials have drawbacks and restrictions of their own, whether related to synthesis, characterization, or integration with desalination systems. However, new developments and studies are always being conducted to find ways around these restrictions. Different materials used in water purification and desalination are demonstrated in **Table 2**.

Material	Pollutant removed	Adsorption capacity (mg/g)	References
Activated carbon fiber (ACF)	NaCl	4.64	[165]
Activated carbon fiber ACF-HNO_3	NaCl	12.8	[166]
ACC-ZnO	NaCl	8.5	[167]
AC-A&AC-C	NaCl	14.-20	[168]
NPC@ACF	NaCl	14.63	[169]
ACNFs	NaCl	827.5	[170]
Neapolitan yellow tuff (NYT)	Humic acid	8.51	[171]
Ca-CLT	NH_3	2.60	[172]
MIL-100(Fe)	Malachite green	485	[173]
MIL-53(Al)–NH_2	Methylene blue, Malachite green	164.9	[174]
ZIF-8	Malachite green	1667	[175]
ZIF-8	Diclofenac sodium	19	[176]
AgOH–MWCNTs	UO_2^{2+}	140	[177]
Fe_3O_4@AMCA-MIL53(Al)	U(VI) and Th(IV)	227.3	[178]
COF-PDAN-AO	UO_2^{2+}	256	[179]
COF-SO_3H	UO_2^{2+}	360	[180]
SCU-COF-1	$^{99}TcO_4^-$	62.8	[181]
COF-S-SH	Hg^{2+}	1350	[182]
COF-BTA-DHBZ	Cr (VI)	384	[135]
COF-LZU8	U (VI)	236	[183]
AS-MWCNTs	Co(II) and Zn(II)	364	[184]
MWCNTs	Trihalomethanes	10.98	[185]
Polyanilline/ MWCNTs	Meloxicam	221.2	[186]
Granular CNTs/ Alumina	Diclofenac sodium and Carbamazepine	106.5	[187]
CNT-FA-SA	Methylene blue	236.5	[188]
Acid treated MWCNTs	Trihalomethanes	0.92–2.41	[189]
Asp-CNT	Malachite green and Methylene blue	637 and 500	[190]

Table 2.
Different adsorbents used in the water purification and desalination application.

3. Summary and outlooks

The existing membrane-based and thermally activated desalination technologies may be replaced by the promising desalination technology known as adsorption. The foundation of adsorption desalination techniques is a porous material. For the application of adsorption, various porous nanomaterials have been studied. The metrics demonstrate that:

1. Although having a relatively modest adsorption capacity, AC and zeolites are safer, less expensive materials with excellent stability.

2. Even though the adsorption capacity and adaptability of graphene and carbon nanotubes are quite great; their safety issues must be appropriately addressed.

3. Because of their high compositional and structural plasticity, MOFs have a high sorption capacity and significant potential; yet cost and stability are crucial factors in advancing applications of MOFs on a large scale.

4. COFs also have a lot of capability for use in adsorption because of their extreme adaptability; however, research on COFs is still in its early stages.

5. The inexpensive cost and large-scale manufacture of other adsorbents like silica, bio-adsorbents, MXenes, fly ash, etc. are advantages, however, their water vapor adsorption capability is insufficient. Findings indicate that MOFs are a sort of adsorbent that shows promise. These porous materials do, however, typically come with a high price tag and issues with stability and hydrolysis. Therefore, the primary difficulty going forward will be the large-scale manufacture of stable, affordable MOFs.

Adsorption desalination is a novel desalination method that has achieved a lot of interest and developed quickly in recent years. Simultaneously, this technology has advanced significantly in theory and experimentation. Adsorption desalination system research, however, is still in the small-scale validation stage and requires a larger-scale demonstration project. Future studies must concentrate on hybrid systems and high-performance adsorbents in order to create more alluring adsorption desalination technologies. If the cost can be further reduced, MOFs, which have a lot of promise in high-performance adsorbent research, will have outstanding practical value. In terms of optimizing the system structure, the thermodynamic performances and heat/mass transfer of adsorbent beds and multifunction mixed systems show greater promise.

Author details

Vishwakarma Ravikumar Ramlal[1] and Savan K. Raj[2]*

1 Department of Chemical Sciences, Ariel University, Ariel, Israel

2 Institut National De Recherche Scientifique (INRS), Energie Materiaux Telecommunications, Quebec, Canada

*Address all correspondence to: savankraj@gmail.com

References

[1] Olabi AG, Abdelkareem MA. Energy storage systems towards 2050. Energy. 2021;**219**:119634. DOI: 10.1016/j.energy.2020.119634

[2] Bigerna S, D'Errico MC, Polinori P. Energy security and RES penetration in a growing decarbonized economy in the era of the 4th industrial revolution. Technological Forecasting and Social Change. 2021;**166**:120648. DOI: 10.1016/j.techfore.2021.120648

[3] Zapata-Sierra A, Cascajares M, Alcayde A, Manzano-Agugliaro F. Worldwide research trends on desalination. Desalination. 2022;**519**:115305. DOI: 10.1016/j.desal.2021.115305

[4] Gittins JR, Hemingway JR, Dajka J-C. How a water-resources crisis highlights social-ecological disconnects. Water Research. 2021;**194**:116937. DOI: 10.1016/j.watres.2021.116937

[5] Tlili I, Alkanhal TA, Othman M, Dara RN, Shafee A. Water management and desalination in KSA view 2030 case study of solar humidification and dehumidification system. Journal of Thermal Analysis and Calorimetry. 2020;**139**:3745-3756. DOI: 10.1007/s10973-019-08700-z

[6] Eliasson J. The rising pressure of global water shortages. Nature. 2015;**517**:6-6. DOI: 10.1038/517006a

[7] Lin S, Zhao H, Zhu L, He T, Chen S, Gao C, et al. Seawater desalination technology and engineering in China: A review. Desalination. 2021;**498**:114728. DOI: 10.1016/j.desal.2020.114728

[8] Sachs J, Moya D, Giarola S, Hawkes A. Clustered spatially and temporally resolved global heat and cooling energy demand in the residential sector. Applied Energy. 2019;**15**(250):48-62. DOI: 10.1016/j.apenergy.2019.05.011

[9] Thu K, Yanagi H, Saha BB, Ng KC. Performance analysis of a low-temperature waste heat-driven adsorption desalination prototype. International Journal of Heat and Mass Transfer. 2013;**65**:662-669. DOI: 10.1016/j.ijheatmasstransfer.2013.06.053

[10] Zhou S, Gong L, Liu X, Shen S. Mathematical modeling and performance analysis for multi-effect evaporation/multi-effect evaporation with thermal vapor compression desalination system. Applied Thermal Engineering. 2019;**159**:113759. DOI: 10.1016/j.applthermaleng.2019.113759

[11] Zarei T, Behyad R. Predicting the water production of a solar seawater greenhouse desalination unit using multi-layer perceptron model. Solar Energy. 2019;**177**:595-603. DOI: 10.1016/j.solener.2018.11.059

[12] Thu K, Chakraborty A, Kim Y-D, Myat A, Saha BB, Ng KC. Numerical simulation and performance investigation of an advanced adsorption desalination cycle. Desalination. 2013;**308**:209-218. DOI: 10.1016/j.desal.2012.04.021

[13] Palenzuela P, Hassan AS, Zaragoza G, Alarcon-Padilla D-C. Steady state model for multi-effect distillation case study: Plataforma solar de Almeria MED pilot plant. Desalination. 2014;**337**:31-42. DOI: 10.1016/j.desal.2013.12.029

[14] Ng KC, Thu K, Kim Y, Chakraborty A, Amy G. Adsorption

desalination: An emerging low-cost thermal desalination method. Desalination. 2013;**308**:161-179. DOI: 10.1016/j.desal.2012.07.030

[15] Alaei Shahmirzadi MA, Hosseini SS, Luo J, Ortiz I. Significance, evolution and recent advances in adsorption technology, materials, and processes for desalination, water softening and salt removal. Journal of Environmental Management. 2018;**215**:324-344. DOI: 10.1016/j.jenvman.2018.03.040

[16] Al-Mousawi FN, Al-Dadah R, Mahmoud S. Novel system for cooling and electricity: Four different integrated adsorption-ORC configurations with two expanders. Energy Conversion and Management. 2017;**152**:72-87. DOI: 10.1016/j.enconman.2017.09.044

[17] Mohammed RH, Rashad E, Huo R, Su M, Chow LC. Pore-size engineered nanoporous silica for efficient adsorption cooling and desalination cycle. npj Clean Water. 2021;**4**:38. DOI: 10.1038/s41545-021-00129-y

[18] Long R, Xia X, Zhao Y, Li S, Liu Z, Liu W. Screening metal-organic frameworks for adsorption-driven osmotic heat engines via grand canonical Monte Carlo simulations and machine learning. Iscience. 2021;**24**:101914. DOI: 10.1016/j.isci.2020.101914

[19] Zhao Y, Luo Z, Long R, Liu Z, Liu W. Performance evaluations of an adsorption-based power and cooling cogeneration system under different operative conditions and working fluids. Energy. 2020;**204**:117993. DOI: 10.1016/j.energy.2020.117993

[20] Mavukkandy MO, Chabib CM, Mustafa I, Al Ghaferi A, AlMarzooqi F. Brine management in desalination industry: From waste to resources generation. Desalination. 2019;**15**(472):114187. DOI: 10.1016/j.desal.2019.114187

[21] Lu GQ, Zhao XS. Nanoporous materials: An overview. Nanoporous Materials: Science and Engineering. 2004;**1**(4):1-2

[22] Yang XY, Chen LH, Li Y, Rooke JC, Sanchez C, Su BL. Hierarchically porous materials: Synthesis strategies and structure design. Chemical Society Reviews. 2017;**46**(2):481-558. DOI: 10.1039/C6CS00829A

[23] Wu J, Xu F, Li S, Ma P, Zhang X, Liu Q, et al. Porous polymers as multifunctional material platforms toward task-specific applications. Advanced Materials. 2019;**31**(4):1802922. DOI: 10.1002/adma.201802922

[24] Thu K, Chakraborty A, Saha BB, Ng KC. Thermo-physical properties of silica gel for adsorption desalination cycle. Applied Thermal Engineering. 2013;**50**(2):1596-1602. DOI: 10.1016/j.applthermaleng.2011.09.038

[25] Raj SK, Sharma V, Yadav A, Indurkar PD, Kulshrestha V. Nano-alumina wrapped carbon microspheres for ultrahigh elimination of pentavalent arsenic and fluoride from potable water. Journal of Industrial and Engineering Chemistry. 2023;**25**(117):402-413. DOI: 10.1016/j.jiec.2022.10.028

[26] Ullah S, Shah SS, Altaf M, Hossain I, El Sayed ME, Kallel M, et al. Activated carbon derived from biomass for wastewater treatment: Synthesis, application and future challenges. Journal of Analytical and Applied Pyrolysis. 2024;**28**:106480. DOI: 10.1016/j.jaap.2024.106480

[27] Díaz B, Sommer-Márquez A, Ordoñez PE, Bastardo-González E, Ricaurte M, Navas-Cárdenas C.

Synthesis methods, properties, and modifications of biochar-based materials for wastewater treatment: A review. Resources. 2024;**13**(1):8. DOI: 10.3390/resources13010008

[28] Miao Z, Che N, Chai W, Zhang M, Wang F. Hierarchical porous carbons derived from porous biomass for high-performance capacitive deionization. The Journal of Physical Chemistry C. 2024. DOI: 10.1021/acs.jpcc.3c07941

[29] Yin L, Wang X, Hu P, Xia H, Liang C, Qu W. Integrated hierarchical porous lignin-based carbon electrode for boosting membrane-free capacitive deionization areal adsorption capacity. International Journal of Biological Macromolecules. 2024;**29**:130065. DOI: 10.1016/j.ijbiomac.2024.130065

[30] Macias-Garcia A, Valenzuela-Calahorro C, Gomez-Serrano V, Espinosa-Mansilla A. Adsorption of Pb2+ by heat-treated and sulfurized activated carbon. Carbon. 1993;**31**(8):1249-1255. DOI: 10.1016/0008-6223(93)90083-M

[31] Goel J, Kadirvelu K, Rajagopal C, Garg VK. Removal of lead (II) by adsorption using treated granular activated carbon: Batch and column studies. Journal of Hazardous Materials. 2005;**125**(1-3):211-220. DOI: 10.1016/j.jhazmat.2005.05.032

[32] Feng W, Borguet E, Vidic RD. Sulfurization of a carbon surface for vapor phase mercury removal–II: Sulfur forms and mercury uptake. Carbon. 2006;**44**(14):2998-3004. DOI: 10.1016/j.carbon.2006.05.053

[33] Skodras G, Orfanoudaki T, Kakaras E, Sakellaropoulos GP. Production of special activated carbon from lignite for environmental purposes. Fuel Processing Technology. 2002;**20**(77):75-87. DOI: 10.1016/S0378-3820(02)00062-0

[34] Raymundo-Piñero E, Cazorla-Amorós D, Linares-Solano A. The role of different nitrogen functional groups on the removal of SO2 from flue gases by N-doped activated carbon powders and fibres. Carbon. 2003;**41**(10):1925-1932. DOI: 10.1016/S0008-6223(03)00180-5

[35] Bhatnagar A, Hogland W, Marques M, Sillanpää M. An overview of the modification methods of activated carbon for its water treatment applications. Chemical Engineering Journal. 2013;**219**:499-511. DOI: 10.1016/j.cej.2012.12.038

[36] Abussaud B, Asmaly HA, Saleh TA, Gupta VK, Atieh MA. Sorption of phenol from waters on activated carbon impregnated with iron oxide, aluminum oxide and titanium oxide. Journal of Molecular Liquids. 2016;**213**:351-359. DOI: 10.1016/j.molliq.2015.08.044

[37] Zhu J, Deng B, Yang J, Gang D. Modifying activated carbon with hybrid ligands for enhancing aqueous mercury removal. Carbon. 2009;**47**(8):2014-2025. DOI: 10.1016/j.carbon.2009.03.047

[38] Wolak E, Orzechowska-Zięba A. Change of the surface and structure of activated carbon as a result of HNO3 modification. Adsorption. 2024;**30**(2):121-128. DOI: 10.1007/s10450-023-00401-2

[39] Naji SZ, Tye CT, Mohamed AR. Role of sulfuric acid modification to coconut shell activated carbon in waste cooking oil upgrading. Biomass Conversion and Biorefinery. 2024;**14**(5):6967-6985. DOI: 10.1007/s13399-022-03018-7

[40] Aggarwal D, Goyal M, Bansal RC. Adsorption of chromium by activated

carbon from aqueous solution. Carbon. 1999;**37**(12):1989-1997. DOI: 10.1016/S0008-6223(99)00072-X

[41] Jia YF, Thomas KM. Adsorption of cadmium ions on oxygen surface sites in activated carbon. Langmuir. 2000;**16**(3):1114-1122. DOI: 10.1021/la990436w

[42] Mouni L, Merabet D, Bouzaza A, Belkhiri L. Adsorption of Pb (II) from aqueous solutions using activated carbon developed from apricot stone. Desalination. 2011;**276**(1-3):148-153. DOI: 10.1016/j.desal.2011.03.038

[43] Liu SX, Chen X, Chen XY, Liu ZF, Wang HL. Activated carbon with excellent chromium(VI) adsorption performance prepared by acid-base surface modification. Journal of Hazardous Materials. 2007;**141**(1):315-319. DOI: 10.1016/j.jhazmat.2006.07.006

[44] Faria PC, Orfao JJ, Pereira MF. Adsorption of anionic and cationic dyes on activated carbons with different surface chemistries. Water Research. 2004;**38**(8):2043-2052. DOI: 10.1016/j.watres.2004.01.034

[45] Jurewicz K, Babeł K, Źiółkowski A, Wachowska H. Ammoxidation of active carbons for improvement of supercapacitor characteristics. Electrochimica Acta. 2003;**48**(11):1491-1498. DOI: 10.1016/S0013-4686(03)00035-5

[46] Lorenc-Grabowska E, Gryglewicz G, Machnikowski J. p-Chlorophenol adsorption on activated carbons with basic surface properties. Applied Surface Science. 2010;**256**(14):4480-4487. DOI: 10.1016/j.apsusc.2010.01.078

[47] Shaarani FW, Hameed BH. Ammonia-modified activated carbon for the adsorption of 2, 4-dichlorophenol. Chemical Engineering Journal. 2011;**169**(1-3):180-185. DOI: 10.1016/j.cej.2011.03.002

[48] Park SJ, Jang YS. Pore structure and surface properties of chemically modified activated carbons for adsorption mechanism and rate of Cr (VI). Journal of Colloid and Interface Science. 2002;**249**(2):458-463. DOI: 10.1006/jcis.2002.8269

[49] Chen W, Cannon FS, Rangel-Mendez JR. Ammonia-tailoring of GAC to enhance perchlorate removal. II: Perchlorate adsorption. Carbon. 2005;**43**(3):581-590. DOI: 10.1016/j.carbon.2004.10.025

[50] Liu QS, Zheng T, Li N, Wang P, Abulikemu G. Modification of bamboo-based activated carbon using microwave radiation and its effects on the adsorption of methylene blue. Applied Surface Science. 2010;**256**:3309-3315. DOI: 10.1016/j.apsusc.2009.12.025

[51] Menéndez JA, Menéndez EM, Iglesias MJ, Garcı'a A, Pis JJ. Modification of the surface chemistry of active carbons by means of microwave-induced treatments. Carbon. 1999:1115-1121. DOI: 10.1016/S0008-6223(98)00302-9

[52] Valente Nabais JM, Carrott PJM, Ribeiro Carrott MML, Menéndez JA. Preparation and modification of activated carbon fibres by microwave heating. Carbon. 2004;**42**:1315-1320. DOI: 10.1016/j.carbon.2004.01.033

[53] Park SJ, Kim BJ. Influence of oxygen plasma treatment on hydrogen chloride removal of activated carbon fibers. Journal of Colloid and Interface Science. 2004;**275**:590595. DOI: 10.1016/j.jcis.2004.03.011

[54] Tanada S, Kawasaki N, Nakamura T, Ohue T, Abe II. Adsorbability of 1, 1,

1, 2-tetrafluoromethane (HFC134a) onto plasma-treated activated carbon in CF4 and CCl4. Journal of Colloid and Interface Science. 1997;**191**:337-340. DOI: 10.1006/jcis.1997.4965

[55] Boudou JP, Martinez-Alonzo A, Tascon JMD. Introduction of acidic groups at the surface of activated carbon by microwave-induced oxygen plasma at low pressure. Carbon. 2000;**38**:1021-1029. DOI: 10.1016/S0008-6223(99)00209-2

[56] Zhang W, Liu HY, Xia QB, Li Z. Enhancement of dibenzothiophene adsorption on activated carbons by surface modification using low temperature oxygen plasma. Chemical Engineering Journal. 2012;**209**:597-600. DOI: 10.1016/j.cej.2012.08.050

[57] Lee KP, Arnot TC, de Mattia D. A review of reverse osmosis membrane materials for desalination—Development to date and future potential. Journal of Membrane Science. 2011;**370**:1-22. DOI: 10.1016/j.memsci.2010.12.036

[58] Zou LD, Morris G, Qi DD. Using activated carbon electrode in electrosorptive deionisation of brackish water. Desalination. 2008;**225**:329-340. DOI: 10.1016/j.desal.2007.07.014

[59] Taher T, Melati EK, Febrina M, Maulana S, Asagabaldan MA, Rianjanu A, et al. Effect of Desilication on Indonesian natural zeolite for the enhancement of ammonium ion removal from aqueous solutions. Silicon. 2024;**16**(3):1309-1319. DOI: 10.1007/s12633-023-02758-z

[60] Masoud MS, Zidan AA, El Zokm GM, Elsamra RM, Okbah MA. Humic acid and nano-zeolite NaX as low cost and eco-friendly adsorbents for removal of Pb (II) and Cd (II) from water: Characterization, kinetics, isotherms and thermodynamic studies. Biomass Conversion and Biorefinery. 2024;**14**(3):3615-3632. DOI: 10.1007/s13399-022-02608-9

[61] Xu J, Yin T, Li Y, Liu N, Shi L, Meng X. Synthesis of high hydrophobicity USY zeolite with excellent VOCs adsorption performance under humid condition: Combined strategy of high temperature steam and acid treatment. Separation and Purification Technology. 2024;**329**:124914. DOI: 10.1016/j.seppur.2023.124914

[62] Wang S, Peng Y. Natural zeolites as effective adsorbents in water and wastewater treatment. Chemical Engineering Journal. 2010;**156**(1):11-24. DOI: 10.1016/j.cej.2009.10.029

[63] Aghakhani A, Mousavi SF, Mostafazadeh-Fard B, Rostamian R, Seraji M. Application of some combined adsorbents to remove salinity parameters from drainage water. Desalination. 2011;**275**(1-3):217-223. DOI: 10.1016/j.desal.2011.03.003

[64] Cinar S, Beler-Baykal B. Ion exchange with natural zeolites: An alternative for water softening? Water Science and Technology. 2005;**51**(11):71-77. DOI: 10.2166/wst.2005.0392

[65] Wajima T, Shimizu T, Yamato T, Ikegami Y. Removal of NaCl from seawater using natural zeolite. Toxicological and Environmental Chemistry. 2010;**92**(1):21-26. DOI: 10.1080/02772240902762958

[66] Wibowo E, Rokhmat M, Murniati R, Abdullah M. Utilization of natural zeolite as sorbent material for seawater desalination. Procedia Engineering. 2017;**170**:8-13. DOI: 10.1016/j.proeng.2017.03.002

[67] Díaz-Nava C, Olguín MT, Solache-Ríos M. Water defluoridation by Mexican heulandite-clinoptilolite. Separation Science and Technology. 2002;**37**(13):3109-3128. DOI: 10.1081/SS-120005662

[68] Samatya S, Yüksel Ü, Yüksel M, Kabay N. Removal of fluoride from water by metal ions (Al3+, La3+, and ZrO2 +) loaded natural zeolite. Separation Science and Technology. 2007;**42**(9):2033-2047. DOI: 10.1080/01496390701310421

[69] Haggerty GM, Bowman RS. Sorption of chromate and other inorganic anions by organo-zeolite. Environmental Science & Technology. 1994;**28**(3):452-458

[70] Ghiaci M, Kia R, Abbaspur A, Seyedeyn-Azad F. Adsorption of chromate by surfactant-modified zeolites and MCM-41 molecular sieve. Separation and Purification Technology. 2004;**40**(3):285-295. DOI: 10.1016/j.seppur.2004.03.009

[71] Armağan B, Özdemir O, Turan M, Çelik MS. The removal of reactive azo dyes by natural and modified zeolites. Journal of Chemical Technology and Biotechnology. 2003;**78**(7):725-732. DOI: 10.1002/jctb.844

[72] Wang SB, Zhu ZH. Characterisation and environmental application of an Australian natural zeolite for basic dye removal from aqueous solution. Journal of Hazardous Materials. 2006;**136**(3):946-952. DOI: 10.1016/j.jhazmat.2006.01.038

[73] Wang SB, Ariyanto E. Competitive adsorption of malachite green and Pb ions on natural zeolite. Journal of Colloid and Interface Science. 2007;**314**(1):25-31. DOI: 10.1016/j.jcis.2007.05.032

[74] Karadag D, Turan M, Akgul E, Tok S, Faki A. Adsorption equilibrium and kinetics of reactive black 5 and reactive red 239 in aqueous solution onto surfactant-modified zeolite. Journal of Chemical & Engineering Data. 2007;**52**:1615-1620. DOI: 10.1021/je7000057

[75] Xue ZT, Li ZL, Ma JH, Bai X, Kang YH, Hao WM, et al. Effective removal of Mg2+ and Ca2+ ions by mesoporous LTA zeolite. Desalination. 2014;**341**:10-18. DOI: 10.1016/j.desal.2014.02.025

[76] Rajput A, Raj SK, Sharma PP, Yadav V, Sarvaia H, Gupta H, et al. Synthesis and characterization of aluminium modified graphene oxide: An approach towards defluoridation of potable water. Journal of Dispersion Science and Technology. 2018. DOI: 10.1080/01932691.2018.1496836

[77] Raj SK, Rajput A, Gupta H, Patidar R, Kulshrestha V. Selective recognition of Fe3+ and Cr3+ in aqueous medium via fluorescence quenching of graphene quantum dots. Journal of Dispersion Science and Technology. 2019;**40**(2):250-255. DOI: 10.1080/01932691.2018.1467775

[78] Yadav V, Raj SK, Rathod NH, Kulshrestha V. Polysulfone/graphene quantum dots composite anion exchange membrane for acid recovery by diffusion dialysis. Journal of Membrane Science. 2020;**611**:118331. DOI: 10.1016/j.memsci.2020.118331

[79] Rajput A, Raj SK, Lebedeva OV, Chesnokova AN, Raskulova TV, Kulshrestha V. Functionalized carbon dots composite cation exchange membranes: Improved electrochemical performance and salt removal efficiency. Colloids and Surfaces A: Physicochemical and Engineering Aspects. 2021;**609**:125677. DOI: 10.1016/j.colsurfa.2020.125677

[80] Raj SK, Sharma J, Kulshrestha V. Facile synthesis of reusable graphene oxide composite magnetic beads for removal of arsenic (III). SPE Polymers. 2021;**2**(1):74-85. DOI: 10.1002/pls2.10031

[81] Raj SK, Yadav V, Bhadu GR, Patidar R, Kumar M, Kulshrestha V. Synthesis of highly fluorescent and water soluble graphene quantum dots for detection of heavy metal ions in aqueous media. Environmental Science and Pollution Research. 2021;**28**:46336-46342. DOI: 10.1007/s11356-020-07891-5

[82] Raj SK, Kulshrestha V. Progress in 2D nanomaterial composites membranes for water purification and desalination. In: 2D Nanomaterials for Energy and Environmental Sustainability. Singapore: Springer Nature Singapore; 2022. pp. 125-148. DOI: 10.1007/978-981-16-8538-5_6

[83] Perreault F, De Faria AF, Elimelech M. Environmental applications of graphene-based nanomaterials. Chemical Society Reviews. 2015;**44**(16):5861-5896. DOI: 10.1039/C5CS00021A

[84] Mishra AK, Ramaprabhu S. Functionalized graphene sheets for arsenic removal and desalination of sea water. Desalination. 2011;**282**:39-45. DOI: 10.1016/j.desal.2011.01.038

[85] Banerjee P, Mukhopadhyay A, Das P. Graphene oxide–nanobentonite composite sieves for enhanced desalination and dye removal. Desalination. 2019;**451**:231-240. DOI: 10.1016/j.desal.2017.06.010

[86] Chen ZH, Liu Z, Hu JQ, Cai QW, Li XY, Wang W, et al. β-Cyclodextrin-modified graphene oxide membranes with large adsorption capacity and high flux for efficient removal of bisphenol A from water. Journal of Membrane Science. 2020;**595**:117510. DOI: 10.1016/j.memsci.2019.117510

[87] Iijima S. Helical microtubules of graphitic carbon. Nature. 1991;**354**(6348):56-58. DOI: 10.1038/354056a0

[88] Pan B, Lin DH, Mashayekhi H, Xing BS. Adsorption and hysteresis of bisphenol A and 17alpha-ethinyl estradiol on carbon nanomaterials. Environmental Science & Technology. 2008;**42**:5480-5485. DOI: 10.1021/es8001184

[89] Avcı A, İnci İ, Baylan N. Adsorption of ciprofloxacin hydrochloride on multiwall carbon nanotube. Journal of Molecular Structure. 2020;**1206**:127711. DOI: 10.1016/j.molstruc.2020.127711

[90] Parlak C, Alver Ö. Adsorption of ibuprofen on silicon decorated fullerenes and single walled carbon nanotubes: A comparative DFT study. Journal of Molecular Structure. 2019;**1184**:110-113. DOI: 10.1016/j.molstruc.2019.02.023

[91] Li HB, Pan LK, Zhang YP, Zou LD, Sun CQ, Zhan YK, et al. Kinetics and thermodynamics study for electrosorption of NaCl onto carbon nanotubes and carbon nanofibers electrodes. Chemical Physics Letters. 2010;**485**(1-3):161166. DOI: 10.1016/j.cplett.2009.12.031

[92] Tofighy MA, Mohammadi T. Salty water desalination using carbon nanotube sheets. Desalination. 2010;**258**(1-3):182-186. DOI: 10.1016/j.desal.2010.03.017

[93] Tofighy MA, Mohammadi T. Permanent hard water softening using carbon nanotube sheets. Desalination. 2011;**2681-3**:208-213. DOI: 10.1016/j.desal.2010.10.028

[94] Yang HY, Han ZJ, Yu SF, Pey KL, Ostrikov K, Karnik R. Carbon

nanotube membranes with ultrahigh specific adsorption capacity for water desalination and purification. Nature Communications. 2013;**4**(1):2220. DOI: 10.1038/ncomms3220

[95] Pendergast MM, Hoek EM. A review of water treatment membrane nanotechnologies. Energy and Environmental Science. 2011;**4**(6):1946-1971. DOI: 10.1039/C0EE00541J

[96] Qiu B, Gorgojo P, Fan X. Adsorption desalination: Advances in porous adsorbents. Chinese Journal of Chemical Engineering. 2022;**42**:151-169. DOI: 10.1016/j.cjche.2021.08.032

[97] Xu Y, Duan F, Cao R, Chen C, Ji W, Li Y, et al. Enhanced salt removal performance using nickel hexacyanoferrate/carbon nanotubes as flow cathode in asymmetric flow electrode capacitive deionization. Desalination. 2023;**566**:116929. DOI: 10.1016/j.desal.2023.116929

[98] Ramlal VR, Patel KB, Raj SK, Srivastava DN, Mandal AK. Self-assembled conjugated coordination polymer nanorings: Role of morphology and redox sites for the alkaline electrocatalytic oxygen evolution reaction. ACS Applied Materials and Interfaces. 2024. DOI: 10.1021/acsami.4c00609

[99] Banglani TH, Chandio I, Ali A, Memon AA, Yang J, Kazi M, et al. Recent Progress in 2D and 3D metal-organic framework-based membranes for water sustainability. Environmental Science: Water Research and Technology. 2024. DOI: 10.1039/D3EW00852E

[100] Xu H, Chen S, Zhao YF, Wang F, Guo F. MOF-based membranes for remediated application of water pollution. ChemPlusChem. 2024;**18**:e202400027. DOI: 10.1002/cplu.202400027

[101] Khatoon N, Ali N, Yang H, Jun W. Review of MOFs and their applications in the preparation of loose nanofiltration membranes for dye and salt fractionation. Desalination and Water Treatment. 2024;**15**:100092. DOI: 10.1016/j.dwt.2024.100092

[102] Sun K, Wei Q, Ji Z, Hampshire S, Dong Y. Structure optimization of ceramic-based metal–organic framework membrane for efficient desalination. Ceramics International. 2024. DOI: 10.1016/j.ceramint.2024.03.363

[103] Yang R, Xu X, Teng J, Zhang Y, Xue Y, Shen M, et al. Porous carbon flow-electrode derived from modified MOF-5 for capacitive deionization. Desalination. 2024;**569**:117077. DOI: 10.1016/j.desal.2023.117077

[104] Fei HH, Rogow DL, Oliver SR. Reversible anion exchange and catalytic properties of two cationic metal-organic frameworks based on Cu(I) and Ag(I). Journal of the American Chemical Society. 2010;**132**(20):7202-7209. DOI: 10.1021/ja102134c

[105] Fei HH, Bresler MR, Oliver SR. A new paradigm for anion trapping in high capacity and selectivity: Crystal-to-crystal transformation of cationic materials. Journal of the American Chemical Society. 2011;**133**(29):11110-11113. DOI: 10.1021/ja204577p

[106] Fei HH, Han CS, Robins JC, Oliver SRJ. A cationic metal–organic solid solution based on Co(II) and Zn(II) for chromate trapping. Chemistry of Materials. 2013;**25**(5):647-652. DOI: 10.1021/cm302585r

[107] Shi PF, Zhao B, Xiong G, Hou YL, Cheng P. Fast capture and separation of, and luminescent probe for, pollutant chromate using a multi-functional

cationic heterometal-organic framework. Chemical Communications. 2012;**48**(66):8231-8233. DOI: 10.1039/C2CC33707J

[108] Zhao X, Bu XH, Wu T, Zheng ST, Wang L, Feng PY. Selective anion exchange with nanogated isoreticular positive metal-organic frameworks. Nature Communications. 2013;**4**:2344. DOI: 10.1038/ncomms3344

[109] Furukawa H, Cordova KE, O'Keeffe M, Yaghi OM. The chemistry and applications of metal-organic frameworks. Science. 2013;**341**:1230444. DOI: 10.1126/science.1230444

[110] Wang Z, Yan T, Chen G, Shi L, Zhang D. High salt removal capacity of metal-organic gel derived porous carbon for capacitive deionization. ACS Sustainable Chemistry & Engineering. 2017;**5**:11637-11644. DOI: 10.1021/acssuschemeng.7b03015

[111] Liu B, Shioyama H, Akita T, Xu Q. Metal-organic framework as a template for porous carbon synthesis. Journal of the American Chemical Society. 2008;**130**(16):5390-5391. DOI: 10.1021/ja7106146

[112] Aijaz A, Sun JK, Pachfule P, Uchida T, Xu Q. From a metal-organic framework to hierarchical high surface-area hollow octahedral carbon cages. Chemical Communications. 2015;**51**(73):13945-13948. DOI: 10.1039/C5CC04230E

[113] Zhao D, Yu G, Ge M, Han M, Meng H, Xiong W, et al. Post-modified linker functionalization of metal organic frameworks for high-efficiency adsorption and quantification of trace organophosphorus pesticides in environmental water. Separation and Purification Technology. 2024;**346**:127470. DOI: 10.1016/j.seppur.2024.127470

[114] Goyal P, Menon D, Jain P, Prakash P, Misra SK. Linker mediated enhancement in reusability and regulation of Pb (II) removal mechanism of Cu-centered MOFs. Separation and Purification Technology. 2023;**318**:123941. DOI: 10.1016/j.seppur.2023.123941

[115] Chen YX, Yuan YM, Yang HY, Wang Q, Ren Y, Guo XH, et al. Hierarchical porous tannic-acid-modified MOFs/alginate particles with synergized adsorption-photocatalysis for water remediation. Separation and Purification Technology. 2024;**330**:125435. DOI: 10.1016/j.seppur.2023.125435

[116] Gao L, Chan KY, Li CY, Xie L, Olorunyomi JF. Highly selective transport of alkali metal ions by nanochannels of polyelectrolyte threaded MIL-53 metal organic framework. Nano Letters. 2019;**19**(8):4990-4996. DOI: 10.1021/acs.nanolett.9b01211

[117] Luo XB, Shen TT, Ding L, Zhong WP, Luo JF, Luo SL. Novel thyminefunctionalized MIL-101 prepared by post-synthesis and enhanced removal of Hg2+ from water. Journal of Hazardous Materials. 2016;**306**:313-322. DOI: 10.1016/j.jhazmat.2015.12.034

[118] Li J, Li XD, Hayat T, Alsaedi A, Chen CL. Screening of zirconium-based metal-organic frameworks for efficient simultaneous removal of antimonite (Sb(III)) and antimonate (Sb(V)) from aqueous solution. ACS Sustainable Chemistry & Engineering. 2017;5(12):11496-11503. DOI: 10.1021/acssuschemeng.7b02720

[119] Raj SK, Carrier AJ, Youden BC, Servos MR, Oakes KD, Zhang X. Electrochemical techniques for uranium extraction from water. Chemical Engineering Journal. 2024:152341. DOI: 10.1016/j.cej.2024.152341

[120] Lim J, Lee H, Lee S, Hong S. Capacitive deionization incorporating a fluidic MOF-CNT electrode for the high selective extraction of lithium. Desalination. 2024:117403. DOI: 10.1016/j.desal.2024.117403

[121] Wei D, Ouyang B, Cao Y, Yan L, Wu B, Chen P, et al. Coordination confined silver-organic framework for high-performance electrochemical deionization. Advanced Science. 2024:2401174. DOI: 10.1002/advs.202401174

[122] Hu Q, Wang D, Liang J, Liu Z, Li J. Porous carbonized N-doped MOF-199 modified with MWCNTs for the deionization of uranium (VI). Separation and Purification Technology. 2024;**330**:125494. DOI: 10.1016/j.seppur.2023.125494

[123] Peng G, Zhang Z, Li H. Designing NiFe-metal organic frameworks@ NiFe-layered double hydroxide/carbon fiber self-supporting electrode for enhanced capacitive deionization. Desalination. 2024:117665. DOI: 10.1016/j.desal.2024.117665

[124] Hayat A, Raza S, Amin MA, Ajmal Z, Alghamdi MM, El-Zahhar AA, et al. Developing new-generation covalent organic frameworks as sustainable catalysts: Synthesis, properties, types and solar energy production. Materials Science and Engineering: R: Reports. 2024;**157**:100771. DOI: 10.1016/j.mser.2024.100771

[125] Xiao L, Wang Z, Guan J. Optimization strategies of covalent organic frameworks and their derivatives for electrocatalytic applications. Advanced Functional Materials. 2024;**34**(11):2310195. DOI: 10.1002/adfm.202310195

[126] Kumar A, Arya K, Mehra S, Kumar A, Mehta SK, Kataria R. Luminescent Zn-MOF@ COF hybrid for selective decontamination of Cu (II) ions and methylene blue dye in aqueous media. Separation and Purification Technology. 2024:126756. DOI: 10.1016/j.seppur.2024.126756

[127] Da Silva VD, Zalewska K, Petrovski Z, Buarque CD, Branco LC, Esteves PM. Covalent organic frameworks as promising materials for the removal of metal and organic pollutants from water. Materials Today Sustainability. 2023;**21**:100279. DOI: 10.1016/j.mtsust.2022.100279

[128] SB MK, Yun YS, Kancharla S. Covalent organic frameworks for critical metal recycling from waste. Coordination Chemistry Reviews. 2024;**507**:215699. DOI: 10.1016/j.ccr.2024.215699

[129] Fang YX, Lin YF, Xu ZL, Mo JW, Li PP. A novel clover-like COFs membrane fabricated via one-step interfacial polymerization for dye/salt separation. Journal of Membrane Science. 2023;**673**:121470. DOI: 10.1016/j.jhazmat.2023.133049

[130] Meng XS, Gui B, Yuan DQ, Zeller M, Wang C. Mechanized azobenzene functionalized zirconium metal-organic framework for on-command cargo release. Science Advances. 2016;**2**(8):e1600480. DOI: 10.1126/sciadv.1600480

[131] Chandrasekara NPGN, Pashley RM. Study of a new process for the efficient regeneration of ion exchange resins. Desalination. 2015;**357**:131-139. DOI: 10.1016/j.desal.2014.11.024

[132] Li J, Yang XD, Bai CY, Tian Y, Li B, Zhang S, et al. A novel benzimidazole-functionalized 2-D COF material:

Synthesis and application as a selective solid-phase extractant for separation of uranium. Journal of Colloid and Interface Science. 2015;**437**:211-218. DOI: 10.1016/j.jcis.2014.09.046

[133] Ding SY, Dong M, Wang YW, Chen YT, Wang HZ, Su CY, et al. Thioether-based fluorescent covalent organic framework for selective detection and facile removal of mercury(II). Journal of the American Chemical Society. 2016;**138**(9):3031-3037. DOI: 10.1021/jacs.5b10754

[134] Li GL, Ye JR, Fang QL, Liu F. Amide-based covalent organic frameworks materials for efficient and recyclable removal of heavy metal lead (II). Chemical Engineering Journal. 2019;**370**:822-830. DOI: 10.1016/j.cej.2019.03.260

[135] Cui FZ, Liang RR, Qi QY, Jiang GF, Zhao X. Efficient removal of Cr(VI) from aqueous solutions by a dual-pore covalent organic framework. Advanced Sustainable Systems. 2019;**3**(4):1800150. DOI: 10.1002/adsu.201800150

[136] Uribe-Romo FJ, Doonan CJ, Furukawa H, Oisaki K, Yaghi OM. Crystalline covalent organic frameworks with hydrazone linkages. Journal of the American Chemical Society. 2011;**133**(30):11478-11481. DOI: 10.1021/ja204728y

[137] Dalapati S, Jin S, Gao J, Xu Y, Nagai A, Jiang D. An azine-linked covalent organic framework. Journal of the American Chemical Society. 2013;**135**(46):17310-17313. DOI: 10.1021/ja4103293

[138] Kuhn P, Antonietti M, Thomas A. Porous, covalent triazine-based frameworks prepared by ionothermal synthesis. Angewandte Chemie, International Edition. 2008;**47**(18):3450-3453

[139] Guo J, Xu Y, Jin S, Chen L, Kaji T, Honsho Y, et al. Conjugated organic framework with three-dimensionally ordered stable structure and delocalized p clouds. Nature Communications. 2013;**4**:2736. DOI: 10.1038/ncomms3736

[140] Segura JL, Mancheño MJ, Zamora F. Covalent organic frameworks based on Schiff-base chemistry: Synthesis, properties and potential applications. Chemical Society Reviews. 2016;**20**:5635-5671. DOI: 10.1039/C5CS00878F

[141] Uribe-Romo FJ, Hunt JR, Furukawa H, Klöck C, O'Keeffe M, Yaghi OM. A crystalline imine-linked 3-D porous covalent organic framework. Journal of the American Chemical Society. 2009;**13**:4570-4571. DOI: 10.1021/ja8096256

[142] Kandambeth S, Shinde DB, Panda MK, Lukose B, Heine T, Banerjee R. Enhancement of chemical stability and crystallinity in porphyrin-containing covalent organic frameworks by intramolecular hydrogen bonds. Angewandte Chemie (International Ed. in English). 2013;**49**:13052-13056. DOI: 10.1002/anie.201306775

[143] Li Y, Wang C, Ma SJ, Zhang HY, Ou JJ, Wei YM, et al. Fabrication of hydrazone-linked covalent organic frameworks using alkyl amine as building block for high adsorption capacity of metal ions. ACS Applied Materials & Interfaces. 2019;**12**:11706-11714. DOI: 10.1021/acsami.8b18502

[144] Wen R, Li Y, Zhang MC, Guo XH, Li X, Li XF, et al. Graphene-synergized 2D covalent organic framework for adsorption: A mutual promotion strategy to achieve stabilization and functionalization simultaneously. Journal of Hazardous Materials. 2018;**358**:273-285. DOI: 10.1016/j.jhazmat.2018.06.059

[145] Wang C, Li R, Xu Y, Ma Z, Qiu Y, Wang C, et al. Effective electrosorption and recovery of phosphorus by capacitive deionization with a covalent organic framework-membrane coating electrode. Desalination. 2024;**570**:117088. DOI: 10.1016/j.desal.2023.117088

[146] Suss ME, Porada S, Sun X, Biesheuvel PM, Yoon J, Presser V. Water desalination via capacitive deionization: What is it and what can we expect from it? Energy & Environmental Science. 2015;**8**:2296-2319. DOI: 10.1039/C5EE00519A

[147] Qi Y, Peng W, Li Y, Zhang F, Fan X. Recent advances in covalent organic frameworks for capacitive deionization: A review. Electrochimica Acta. 2024:143870. DOI: 10.1016/j.electacta.2024.143870

[148] Bhatnagar A, Sillanpää M. Utilization of agro-industrial and municipal waste materials as potential adsorbents for water treatment—A review. Chemical Engineering Journal. 2010;**157**:277-296. DOI: 10.1016/j.cej.2010.01.007

[149] Sonqishe GBTM, Iwouha E, Petrik L. Treatment of Brines Using Commercial Zeolites and Zeolites Synthesized from Fly Ash Derivative. The International Mine Water Conference Proceedings; 2008. Available from: http://hdl.handle.net/11394/8905

[150] Raj SK, Sharma V, Srivastava DN, Kulshrestha V. In-situ evolution of bimetallic Fe/Ni/Co nanohybrids on MXene for improved electrocatalytic oxygen evolution reaction. International Journal of Hydrogen Energy. 2023;**48**(96):37732-37745. DOI: 10.1016/j.ijhydene.2022.11.093

[151] Raj SK, Patel KB, Sharma V, Srivastava DN, Kulshrestha V. Nanopalladium-anchored MXene Nanoflowers for boosting Electrocatalytic hydrogen evolution reaction. Energy and Fuels. 2023;**37**(21):16856-16865. DOI: 10.1021/acs.energyfuels.3c03245

[152] Indurkar PD, Raj SK, Kulshrestha V. Multivariate modeling and process optimization of Hg (II) remediation using solvothermal synthesized 2D MX/Fe3O4 by response surface methodology: Characteristics and mechanism study. Environmental Science and Pollution Research. 2023:76085-76103. DOI: 10.1007/s11356-023-27687-7

[153] Raj SK, Bhadu GR, Upadhyay P, Kulshrestha V. Three-dimensional Ni/Fe doped graphene oxide@ MXene architecture as an efficient water splitting electrocatalyst. International Journal of Hydrogen Energy. 2022;**47**(99):41772-41782. DOI: 10.1016/j.ijhydene.2022.05.204

[154] Raj SK, Kirti SV, Srivastava DN, Kulshrestha V. Single-step synthesis of well-ordered hierarchical nickel nanostructures for boosting the oxygen evolution reaction. Energy and Fuels. 2022;**36**(22):13786-13795. DOI: 10.1021/acs.energyfuels.2c03097

[155] Ihsanullah I. MXenes (two-dimensional metal carbides) as emerging nanomaterials for water purification: Progress, challenges and prospects. Chemical Engineering Journal. 2020;**388**:124340. DOI: 10.1016/j.cej.2020.124340

[156] Zhang TQ, Hao S, Zhao JK, Jia ZQ, Tan HW, Yang Y, et al. Exfoliated MXene/poly-melamine-formaldehyde composite membranes for removal of heavy metals and organics from aqueous solutions. Journal of Hazardous Materials. 2024;**463**:132866. DOI: 10.1016/j.jhazmat.2023.132866

[157] Al-Ghouti MA, Al Disi ZA, Al-Kaabi N, Khraisheh M. Mechanistic insights into the remediation of bromide ions from desalinated water using roasted date pits. Chemical Engineering Journal. 2017;**308**:463-475. DOI: 10.1016/j.cej.2016.09.091

[158] Hettiarachchi E, Perera R, Chandani Perera AD, Kottegoda N. Activated coconut coir for removal of sodium and magnesium ions from saline water. Desalination and Water Treatment. 2016;**57**(47):22341-22352. DOI: 10.1080/19443994.2015.1129649

[159] Li Y, Zhao M, Yan H, Chen Y, Liu Y, Jiang H, et al. Silica from rice husk for sludge-based biochar modification: As a novel adsorbent for lead. Journal of Water Process Engineering. 2024;**60**:105218. DOI: 10.1016/j.jwpe.2024.105218

[160] Amor TB, Lebeau B, Tlili M, Amor MB. Synthesis, characterization and application of glucamine-modified mesoporous silica type SBA-15 for the removal of boron from natural water. Desalination. 2014;**351**:82-87. DOI: 10.1016/j.desal.2014.07.028

[161] Tang YP, Chung TS, Weber M, Maletzko C. Development of novel diol-functionalized silica particles toward fast and efficient boron removal. Industrial and Engineering Chemistry Research. 2017;**56**(40):11618-11627. DOI: 10.1021/acs.iecr.7b03115

[162] Xu L, Liu Y, Hu H, Wu Z, Chen Q. Synthesis, characterization and application of a novel silica based adsorbent for boron removal. Desalination. 2012;**294**:1-7. DOI: 10.1016/j.desal.2012.02.030

[163] Ali ES, Askalany AA, Harby K, Diab MR, Alsaman AS. Adsorption desalination-cooling system employing copper sulfate driven by low grade heat sources. Applied Thermal Engineering. 2018;**136**:169-176. DOI: 10.1016/j.applthermaleng.2018.03.014

[164] Anis SF, Hashaikeh R, Hilal N. Functional materials in desalination: A review. Desalination. 2019;**468**:114077. DOI: 10.1016/j.desal.2019.114077

[165] Wang G, Pan C, Wang L, Dong Q, Yu C, Zhao Z, et al. Activated carbon nanofiber webs made by electrospinning for capacitive deionization. Electrochimica Acta. 2012;**69**:65-70. DOI: 10.1016/j.electacta.2012.02.066

[166] Wu T, Wang G, Dong Q, Qian B, Meng Y, Qiu J. Asymmetric capacitive deionization utilizing nitric acid treated activated carbon fiber as the cathode. Electrochimica Acta. 2015;**176**:426-433. DOI: 10.1016/j.electacta.2015.07.037

[167] Myint MT, Al-Harthi SH, Dutta J. Brackish water desalination by capacitive deionization using zinc oxide micro/nanostructures grafted on activated carbon cloth electrodes. Desalination. 2014;**344**:236-242. DOI: 10.1016/j.desal.2014.03.037

[168] Xu Y, Duan F, Li Y, Cao H, Chang J, Pang H, et al. Enhanced desalination performance in asymmetric flow electrode capacitive deionization with nickel hexacyanoferrate and activated carbon electrodes. Desalination. 2021;**514**:115172. DOI: 10.1016/j.desal.2021.115172

[169] Men L, Chen C, Liu A, Yu S, Zhou J, Xie Y, et al. N-doped porous carbon-based capacitive deionization electrode materials loaded with activated carbon fiber for water desalination applications. Journal of Environmental Chemical Engineering. 2022;**10**(3):107943. DOI: 10.1016/j.jece.2022.107943

[170] Sokolowski K, Blazewicz S, Nocun M, Fraczek-Szczypta A. Carbon micro-and nanofibrous materials with high adsorption capacity for water desalination. Desalination. 2021;**503**:114936. DOI: 10.1016/j.desal.2021.114936

[171] Capasso S, Salvestrini S, Coppola E, Buondonno A, Colella C. Sorption of humic acid on zeolitic tuff: A preliminary investigation. Applied Clay Science. 2005;**28**(1-4):159-165. DOI: 10.1016/j.clay.2004.01.010

[172] Du Q, Liu S, Cao Z, Wang Y. Ammonia removal from aqueous solution using natural Chinese clinoptilolite. Separation and Purification Technology. 2005;**44**(3):229-234. DOI: 10.1016/j.seppur.2004.04.011

[173] Huo SH, Yan XP. Metal–organic framework MIL-100 (Fe) for the adsorption of malachite green from aqueous solution. Journal of Materials Chemistry. 2012;**22**(15):7449-7455

[174] Li C, Xiong Z, Zhang J, Wu C. The strengthening role of the amino group in metal–organic framework MIL-53 (Al) for methylene blue and malachite green dye adsorption. Journal of Chemical and Engineering Data. 2015;**60**(11):3414-3422. DOI: 10.1021/acs.jced.5b00692

[175] Abdi J, Vossoughi M, Mahmoodi NM, Alemzadeh I. Synthesis of metal-organic framework hybrid nanocomposites based on GO and CNT with high adsorption capacity for dye removal. Chemical Engineering Journal. 2017;**326**:1145-1158. DOI: 10.1016/j.cej.2017.06.054

[176] Bhadra BN, Ahmed I, Kim S, Jhung SH. Adsorptive removal of ibuprofen and diclofenac from water using metal-organic framework-derived porous carbon. Chemical Engineering Journal. 2017;**314**:50-58. DOI: 10.1016/j.cej.2016.12.127

[177] Zare F, Ghaedi M, Daneshfar A, Agarwal S, Tyagi I, Saleh TA, et al. Efficient removal of radioactive uranium from solvent phase using AgOH–MWCNTs nanoparticles: Kinetic and thermodynamic study. Chemical Engineering Journal. 2015;**273**:296-306. DOI: 10.1016/j.cej.2015.03.002

[178] Alqadami AA, Naushad M, Alothman ZA, Ghfar AA. Novel metal–organic framework (MOF) based composite material for the sequestration of U (VI) and Th (IV) metal ions from aqueous environment. ACS Applied Materials and Interfaces. 2017;**9**(41):36026-36037. DOI: 10.1021/acsami.7b10768

[179] Li FF, Cui WR, Jiang W, Zhang CR, Liang RP, Qiu JD. Stable sp2 carbon-conjugated covalent organic framework for detection and efficient adsorption of uranium from radioactive wastewater. Journal of Hazardous Materials. 2020;**392**:122333. DOI: 10.1016/j.jhazmat.2020.122333

[180] Xiong XH, Yu ZW, Gong LL, Tao Y, Gao Z, Wang L, et al. Ammoniating covalent organic framework (COF) for high-performance and selective extraction of toxic and radioactive uranium ions. Advanced Science. 2019;**6**(16):1900547. DOI: 10.1002/advs.201900547

[181] He L, Liu S, Chen L, Dai X, Li J, Zhang M, et al. Mechanism unravelling for ultrafast and selective 99 TcO 4– uptake by a radiation-resistant cationic covalent organic framework: A combined radiological experiment and molecular dynamics simulation study. Chemical Science. 2019;**10**(15):4293-4305. DOI: 10.1039/C9SC00172G

[182] Sun Q, Aguila B, Perman J, Earl LD, Abney CW, Cheng Y, et al. Postsynthetically modified covalent organic frameworks for efficient and effective mercury removal. Journal of the American Chemical Society. 2017;**139**(7):2786-2793. DOI: 10.1021/jacs.6b12885

[183] Dai Li Z, Zhang HQ, Xiong XH, Luo F. U (VI) adsorption onto covalent organic frameworks-TpPa-1. Journal of Solid State Chemistry. 2019;**1**(277):484-492. DOI: 10.1016/j.jssc.2019.06.044

[184] Egbosiuba TC, Abdulkareem AS. Highly efficient as-synthesized and oxidized multi-walled carbon nanotubes for copper (II) and zinc (II) ion adsorption in a batch and fixed-bed process. Journal of Materials Research and Technology. 2021;**15**:2848-2872. DOI: 10.1016/j.jmrt.2021.09.094

[185] Dehghani MH, Mohammadi M, Mohammadi MA, Mahvi AH, Yetilmezsoy K, Bhatnagar A, et al. Equilibrium and kinetic studies of trihalomethanes adsorption onto multi-walled carbon nanotubes. Water, Air, and Soil Pollution. 2016;**227**:1-7. DOI: 10.1007/s11270-016-3029-2

[186] Dutra FV, Pires BC, Nascimento TA, Borges KB. Functional polyaniline/multiwalled carbon nanotube composite as an efficient adsorbent material for removing pharmaceuticals from aqueous media. Journal of Environmental Management. 2018;**221**:28-37. DOI: 10.1016/j.jenvman.2018.05.051

[187] Wei H, Deng S, Huang Q, Nie Y, Wang B, Huang J, et al. Regenerable granular carbon nanotubes/alumina hybrid adsorbents for diclofenac sodium and carbamazepine removal from aqueous solution. Water Research. 2013;**47**(12):4139-4147. DOI: 10.1016/j.watres.2012.11.062

[188] Lei Y, Huang Q, Dou J, Huang H, Yang G, Deng F, et al. Fast adsorptive removal of cationic organic dye by anionic group functionalized carbon nanotubes with high efficiency. Colloid and Interface Science Communications. 2021;**40**:100328. DOI: 10.1016/j.colcom.2020.100328

[189] Lu C, Chung YL, Chang KF. Adsorption of trihalomethanes from water with carbon nanotubes. Water Research. 2005;**39**(6):1183-1189. DOI: 10.1016/j.watres.2004.12.033

[190] Saxena M, Lochab A, Saxena R. Asparagine functionalized MWCNTs for adsorptive removal of hazardous cationic dyes: Exploring kinetics, isotherm and mechanism. Surfaces and Interfaces. 2021;**25**:101187. DOI: 10.1016/j.surfin.2021.101187

Chapter 6

Machine Learning Assisting Intelligent Control of Evaporation Performance in Porous Media

Hui Wang, Shaoxuan Xu and Zhiguo Qu

Abstract

In this chapter, four parts about how machine learning can assist intelligent control of evaporation performance in porous media are introduced: first, the two–/three-dimensional porous media structure information is restructured by machine learning; second, the coefficient relationships between porous media structure and transport performance are distinguished and built by machine learning; third, heat and mass transfer fields in porous media are predicting by machine learning; fourth, the best structure information of porous media can be restructured by machine learning. Finally, the conclusion and respective about the machine learning assisting intelligent control of evaporation performance in porous media in the future are introduced.

Keywords: machine learning, evaporation performance, porous media, heat and mass transfer, structure information

1. Introduction

Evaporation is an important step in the desalination process. In order to expand the evaporation surface area, porous media is widely used. The evaporation performance of porous materials refers to the efficiency and effectiveness with which these materials can facilitate the process of liquid evaporation in **Figure 1**.

The efficiency of evaporation depends significantly on the complex microstructures of porous media, which govern the heat and mass transfer properties [1, 2]. The intricate interplay between the material's pore size, distribution, and connectivity creates challenges in predicting and optimizing evaporation performance. Traditional experimental methods and theoretical models often fall short in providing accurate predictions due to the complexity and variability inherent in porous structures. First, generating accurate and representative porous media structures is fundamental for understanding and optimizing their performance. Generating a large number of porous structures is crucial because training datasets for field or property prediction require a significant amount of such structures. However, generating the required complex porous structures using traditional numerical simulation methods can be

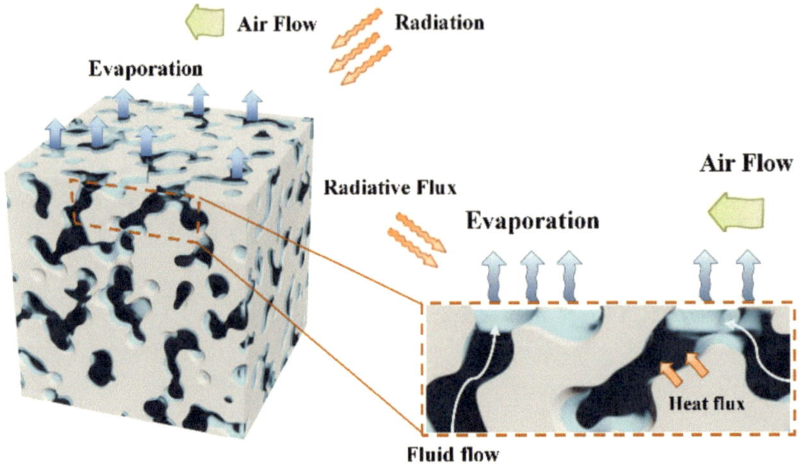

Figure 1.
Evaporation process in porous media.

challenging and computationally expensive [3]. This is because these methods often fall short in accurately replicating the intricate geometries and material properties of natural porous structures [4]. Techniques, such as X-ray microtomography, can reproduce specific existing materials but are costly and not always available [5]. For instance, Bultreys et al. [6] enhanced laboratory micro-computed tomography (micro-CT) technology to enable real-time observation of dynamic processes in geological materials at the pore scale, such as fluid flow and chemical reactions. By utilizing a liquid metal jet X-ray source, high-performance detectors, and a rapid data acquisition system, alongside developing intelligent reconstruction and four-dimensional (4D) image analysis algorithms, the researchers successfully conducted experiments with a time resolution of 12 seconds per scan, although the image quality was inferior to that of synchrotron imaging. This method was validated in two-phase flow experiments in sandstone and solute transport experiments in limestone, revealing crucial details of pore-filling events and solute transport. However, laboratory CT technology still faces some limitations, such as low time and spatial resolution due to the limited flux of X-ray sources.

Besides, previous works stated that the evaporation performance of porous materials is correlated to their microstructure features. For instance, Fei et al. [7] studied the impact of different levels of contact angles and pore sizes on evaporation dynamics. They proposed a strategy that leverages the heterogeneity of contact angles and pore sizes to extend the constant rate period and enhance the evaporation rate. The relationship between the structural characteristics of porous media and their evaporation performance is complex and multifaceted. Identifying and modeling these relationships are essential for optimizing material design. Varied traditional methods have been employed to elucidate the complex relationship between structural characteristics and the evaporation performance of porous media. However, most of these approaches, often rooted in experimental techniques and numerical models [8, 9], have provided foundational insights into the evaporation performance of porous structures. For instance, Paap et al. [10] designed an experimental apparatus to measure the mass balance of water evaporation through porous media. Three scenarios were used to observe the impact of different wettabilities on evaporation:

pure hydrophilic sand, a layer of hydrophobic sand covering hydrophilic sand, and hydrophilic sand mixed with hydrophobic sand. However, the drawback of traditional methods is that they may not fully capture the complexity of multi-scale pore structures, leading to limitations in interpreting heat and mass transport properties. Additionally, these methods often require extensive experimental data and computational resources, making them difficult to apply to large-scale or complex porous media.

The heat and mass transfer processes significantly impact the evaporation in porous media. Understanding these fields within porous media during evaporation is essential for optimizing performance in practical applications. Traditional simulation methods include experimental measurements [11, 12], numerical simulation [13], lattice Boltzmann method [14], and finite-Volume method [15]. Experimental measurement is a commonly used method to characterize heat transfer, fluid dynamics, and substance movement within porous materials. Wang et al. [16] conducted experimental simulations of gas flow in porous media and measured gas production data under various temperatures and pressures to quantify the surface diffusion effect of adsorbed gases. They proposed a novel experimental method capable of measuring dynamic gas mass transfer during pressure decline and developed a mathematical model that incorporates surface diffusion, thereby distinguishing different stages of gas transport. Indeed, the experimental sample size is constrained, making it difficult to extrapolate fine-scale experimental outcomes to broader application domains. Moreover, the experimental measurement of heat, flow, and mass transfer processes in porous media presents significant challenges for visualization.

Therefore, direct numerical simulations [e.g., molecular dynamics (MD) simulations, lattice Boltzmann method (LBM)] are widely utilized for predicting heat, flow, and mass transfer in porous media. For instance, Hong et al. [17] used molecular dynamics simulations to construct jennite models with varying porosity using the ClayFF force field and calculated the thermal conductivity using the Green-Kubo and Müller-Plathe methods. The results indicated that the thermal conductivity of jennite significantly decreases with increasing porosity, consistent with predictions from the coherent potential model. Although molecular dynamics (MD) simulations can precisely predict heat and mass transport in porous media, they are computationally expensive and restricted to problems with simple geometries at the nanoscale [18]. To overcome the limitations related to size, LBM is employed. For instance, Yoshino and Inamuro [19] used LBM to simulate the flow, heat transfer, and mass transfer in 3D porous materials. Compared to the experimental data, the proposed method demonstrated good accuracy. However, the computational complexity and time consumption of the LBM increase with the size of the simulation domain, which makes them unavailable for studying porous materials with large domains [20].

Recently, deep learning has been extensively employed in various fields, which can be used to predict the structure information of porous media. By integrating physical principles with ML techniques, we can develop models that not only provide quick predictions but also maintain high levels of accuracy. Researchers used deep learning methods to predict heat and mass transfer properties and fields are represented in section four.

Once the relationships between structure and performance are established and the transfer fields can be accurately predicted, the next step is to optimize the porous media structures for enhanced evaporation performance.

In summary, this chapter will demonstrate how machine learning techniques can revolutionize the design, analysis, and optimization of porous media for evaporation control, offering substantial improvements over traditional methods and paving the way for more efficient and effective applications in various fields.

2. Reconstruction of porous media structure by machine learning

Employing machine learning to generate porous structure samples emerges as a promising alternative. Machine learning excels in identifying patterns and features from large datasets, making them well-suited for the complex task of generating porous materials with specific properties [21]. Generative models such as Generative Adversarial networks (GANs) and Variational Autoencoders (VAEs) have demonstrated significant potential in synthesizing realistic material structures by learning from high-dimensional data distributions. GANs, in particular, have been successfully applied to generate three-dimensional pore-space realizations, capturing both the statistical and geometric properties of the training datasets.

This section aims to provide a comprehensive overview of state-of-the-art machine learning methods for generating porous materials. We discuss the generation of 2D and 3D porous structures separately, covering methods such as Autoencoders, Generative Adversarial Networks, and Diffusion models. We evaluate their effectiveness in capturing the complex geometries and functionalities of porous structures and discuss their limitations and future prospects. By leveraging advancements in machine learning, this section highlights the potential of these technologies to revolutionize the design and characterization of porous materials, paving the way for innovative applications in various scientific and engineering domains.

2.1 Generation of 2D porous media structures

Machine learning methods used to generate irregular porous materials are introduced. Due to the high computational cost of generating 3D structures, early research mainly focused on using machine learning methods to construct 2D porous structures. There are two types of machine learning methods for generating porous structures, one is supervised learning and the other one is unsupervised learning. In supervised learning, a model is trained using labeled data, where each input (such as a pixel in an image) is associated with a corresponding output (such as a phase value) [22]. The model is trained to recognize patterns in the images. For each pixel, the model learns to predict its phase value by considering the phase values of its neighboring pixels. New images are generated layer by layer. The trained model predicts the phase values for pixels in a new layer based on the known phases in adjacent layers. Supervised learning methods often achieve high accuracy because they leverage labeled data, allowing the model to learn precise relationships between inputs and outputs [23]. However, the need for a large, well-labeled dataset can be a significant limitation, as acquiring and labeling such data can be time-consuming and expensive. The model may become too specialized for the training data, performing well on known data but poorly on unseen data or different distributions. This is not useful for generating a new optimized structure which has not been seen before. Due to this limitation, only unsupervised learning methods are introduced in this section.

Variational Autoencoders (VAEs) represent a significant approach in microstructure reconstruction. Laloy et al. [24] introduced VAECNN, a VAE model incorporating convolutional layers to reconstruct complex binary porous media. The structure

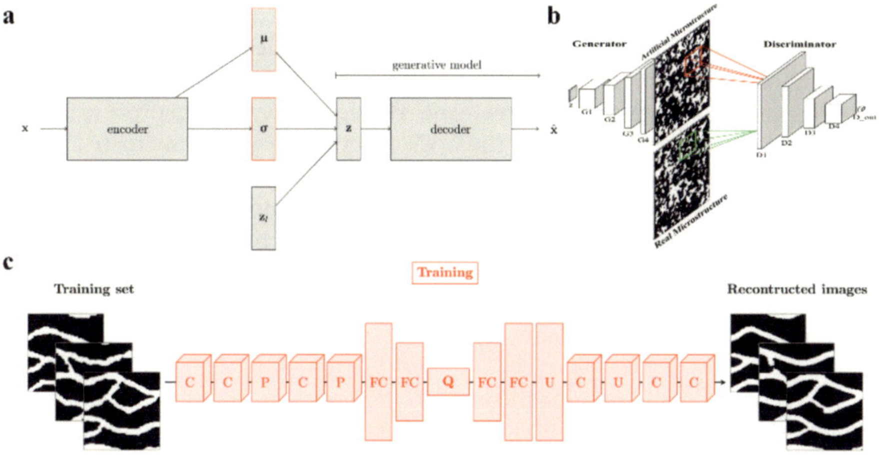

Figure 2.
Architectures of generative models. (a) Structure of Variational Autoencoder. Reprinted from Ref. [24]. (b) Architecture of GAN from the work of Yang et al. Reprinted from Ref. [25]. (c) Represent of complete architecture of Variational Autoencoder proposed by Laloy et al. Reprinted from Ref. [24].

of a VAE is represented in **Figure 2(a)** and the architecture of the proposed VAECNN is shown in **Figure 2(b)**. Utilizing a variational autoencoder (VAE)-type deep neural network, one can construct a low-dimensional parametric representation, facilitating random sampling from a standard normal distribution that is uncorrelated to generate instances consistent with the training set. This technique proves superior to other dimensionality reduction methods like principal component analysis (PCA), optimization-PCA (OPCA), and discrete cosine transform (DCT) in the context of unconditional geological statistical simulation for channelized prior models. Although it necessitates numerous training samples, this method proves more apt for probabilistic or deterministic inversion instead of unconditional or point-conditioned geostatistical simulation. The success of dimension-reduction-based inversion is evidenced by examples of 2D steady-state and 3D transient hydraulic tomography data inversion, surpassing the current top multipoint statistical inversion method, SGR, in the 2D case study.

Generative Adversarial networks (GANs) are another type of generative model used and developed for porous structure reconstruction. By training on real data, GANs learn to produce highly realistic textures and structures, which is crucial for accurately simulating the complex morphology of porous materials. For instance, Yang et al. [25] proposed a deep adversarial learning methodology for microstructural material design using generative adversarial networks (GANs) to identify latent variables of microstructures. This method models the connection between latent variables and microstructure images, utilizing physical simulations to assess material properties. The structure of the proposed model is illustrated in **Figure 2(c)**. To reduce computational expenses, a GP-Hedge Bayesian optimization framework is employed to fine-tune the microstructure for desired properties. This technique offers a comprehensive solution, embedding microstructures into a low-dimensional, nonlinear space without losing information. GANs effectively capture intricate microstructural features and provide scalability for transforming microstructures into low-dimensional representations. The discriminator can be repurposed to develop predictive models for structure-property relationships.

Yang et al. used a dataset of 5000 synthetic microstructure images generated with the Gaussian Random Field (GRF) method. The GAN model was validated by comparing original and generated microstructures using two-point and lineal-path correlation functions, showing good agreement. The design evaluation process involves propagating latent variables through the GAN generator to obtain microstructure images and using physics-based simulations for material properties. For optimization, the Bayesian framework balances exploration and exploitation. The methodology demonstrates scalability by generating arbitrary-sized microstructures and improves predictive model accuracy through transfer learning.

2.2 Generation of 3D porous media structures

The structure of porous media is complex, most of them have distinct morphologies, such as fibrous structures and helical structures, that cannot be adequately generated using simple CNNs. This is where GANs, particularly conditional GANs (cGANs), come into play. For instance, Feng et al. [26] introduced a deep learning approach to expedite the reconstruction of three-dimensional porous media from two-dimensional images. They used a cGAN to describe the relationship between sampled images (containing conditional data) and target pore-solid images. They chose to use U-Net as the architecture of the generator of GAN. Compared to traditional multipoint statistics (MPS) methods, this approach can instantly transform sampled images into target images, significantly improving reconstruction speed—about 760 times faster for 2D reconstruction and approximately 25 times faster for 3D reconstruction. The researchers validated their approach using sandstone samples, comparing the reconstructed microstructure to the target using several quantitative measures. These included the two-point correlation function, lineal-path function, two-point cluster function, and local porosity distribution. The comparison indicated that the proposed method not only maintains high efficiency but also preserves good agreement with the statistical functions of the target system. However, this method relies on a training dataset, which may affect its generalization ability for complex samples. This study opens new avenues for addressing computationally intensive problems by integrating machine learning/deep learning techniques.

Given the high training costs of GANs and the challenges in generating detailed structures, many researchers have sought to integrate GANs with other techniques to enhance structural detail and reduce computational expenses. Reza et al. [27] introduced an innovative technique for reconstructing 3D multi-scale porous media from single-scale images through the application of machine learning. Initially, the study showcased the ability of a GAN to reconstruct Berea sandstone. Building on this foundation, they devised a combined approach integrating GANs with autoencoder (AE) neural networks to reconstruct multi-scale sandy microstructures. The architectures of the proposed GAN and AE are depicted in **Figure 3(a)** and **(b)**. The GAN was employed for the volumetric reconstruction of 3D media featuring inter-grain porosity, whereas the AE network added intra-grain porosity. This technique was further tested by reconstructing multi-scale carbonate samples from an Iranian oilfield. In their approach, GANs were trained with a dataset containing inter-grain pore spaces to generate 3D images. Following this, a sparse AE was trained to create intra-grain pores from the GAN output, yielding high-resolution images that incorporate multi-scale pore structures. Additionally, the AE was used for the 2D multi-scale reconstruction of carbonate samples, effectively generating outputs that include both inter- and intra-grain pore spaces. The coupled GAN and AE reconstruction results

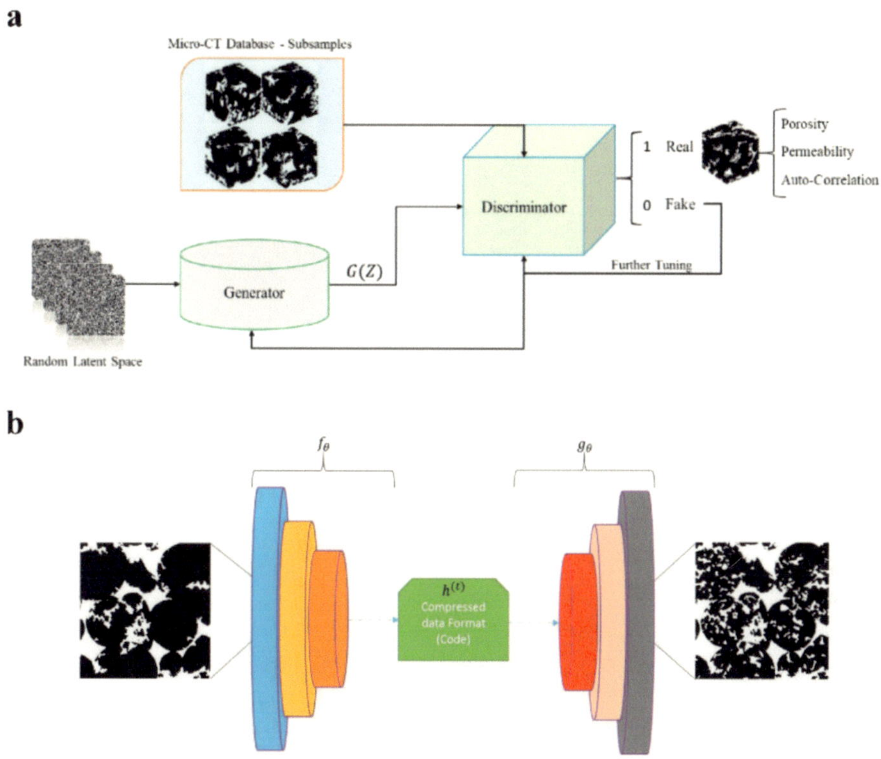

Figure 3.
Architectures of generative models. (a) Architecture of GAN from the work of Reza et al. (b) Architecture of AE from the work of Reza et al. Reprinted from Ref. [27].

showed better consistency, indicating a more accurate reconstruction of the medium's structure, including long-range connectivity between pores.

VAEs and GANs have broad application in fields such as microstructure reconstruction, but each has its own set of limitation. VAEs use maximum likelihood estimation to approximate the gap between the mixture of Gaussian distributions and the real data distribution, which can result in lack of structure details and generate blurry images. GANs may repeatedly generate the same image, a phenomenon known as mode collapse, which limits their ability to produce diverse outputs. Also, due to the adversarial nature of GANs loss function, the training process can be unstable. These limitations hinder the further development of GANs in practical applications. In contrast, diffusion models in recent years have exhibited exceptional performance in the domain of Artificial Intelligence Generated Content (AIGC). These models not only possess advantages in computational efficiency, expandability, and parallelism but also excel in generating high-quality and diverse content. Lyu et al. [28] employed diffusion probabilistic models for the reconstruction of both 2D and 3D random porous material microstructures. The structure of the proposed diffusion model is based on 3D U-Net, the architecture of 3D U-Net is shown in **Figure 4(a)**. They proposed a novel approach based on Denoising Diffusion Probabilistic Models (DDPM) and Deterministic Diffusion Implicit Models (DDIM). DDPM simulates the forward diffusion process of an image, while the DDIM recovers the image through continuous interpolation in the latent space, effectively regulating the randomness of the microstructures and creating gradient materials. The forward noising process

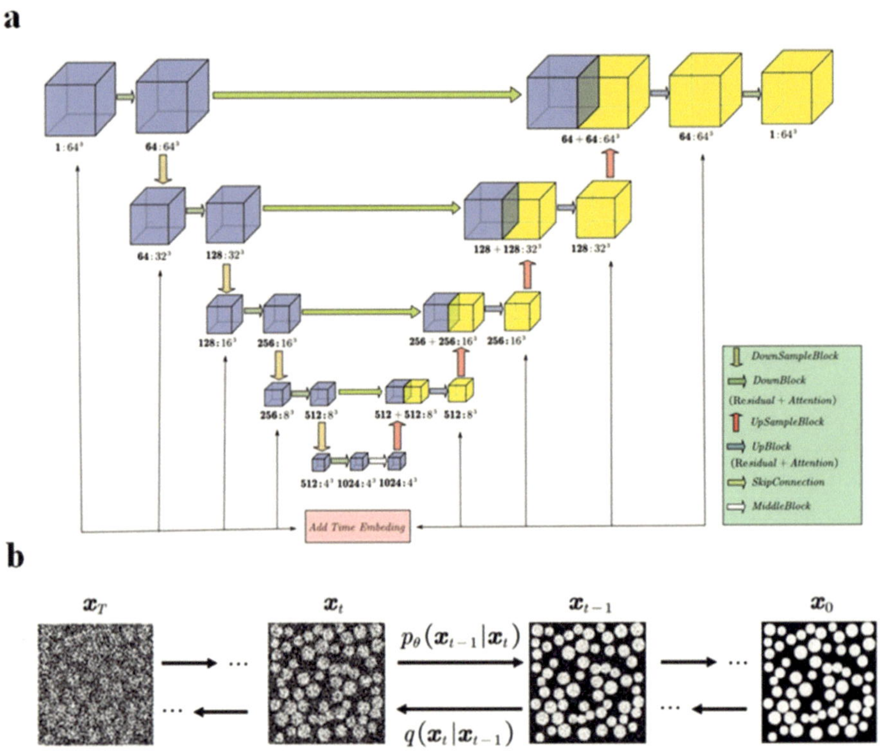

Figure 4.
Architectures of generative models. (a) Represent of architecture of 3D U-Net in diffusion model. (b) The forward noising process and reverse denoising process in diffusion model. Reprinted from Ref. [28].

and reverse denoising process in the diffusion model are represented in **Figure 4(b)**. Additionally, they extended this method to a three-dimensional framework, using permeability feature encoding to achieve the conditional generation of three-dimensional irregular porous structures that adhere to a defined permeability range.

Figure 5 represents the comparison between original porous structures and generated porous structures. The consistency between the generated microstructures and the original structures was evaluated utilizing statistical metrics like the Fourier descriptors, linear path function, and two-point correlation function. This work provides new tools for materials inverse design, demonstrating high efficiency in precise reconstruction and the regulation of microstructure randomness. Despite the complex computations involved, DDPM and DDIM offer a more effective approach to handling high-dimensional data.

Looking forward, advancing the field of generating porous materials entails enhancing generative models like VAEs, GANs, DDPM, and DDIM to accurately simulate complex geometries and functionalities while addressing challenges such as data requirements, training complexities, and computational costs. Future efforts should focus on augmenting and synthesizing diverse datasets, integrating physics-based simulations for realistic material behavior prediction, and extending capabilities to model multi-scale porous structures. Automation and scalability improvements are crucial for efficiently generating large-scale structures, alongside robust validation and uncertainty quantification methods to ensure reliability in various applications. Interdisciplinary collaboration between material scientists, engineers, and data

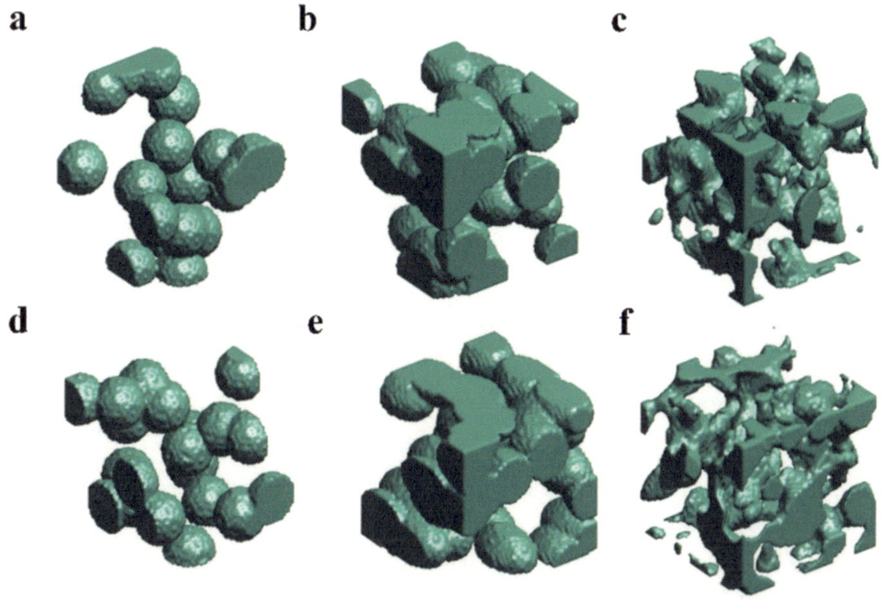

Figure 5.
Comparison between original three-dimensional microstructure and reconstructed microstructure of porous materials; (a) Original spherical structure. (b) Original ellipsoidal structure. (c) Original irregular porous structure. (d) Generated spherical structure. (e) Generated ellipsoidal structure. (f) Irregular porous structure. Reprinted from Ref. [28].

scientists will be essential for leveraging expertise and pushing the boundaries of generative modeling in porous materials research.

3. The coefficient relationships between porous media structure and transport performance

Studying the relationship between evaporation performance and structure is crucial for optimizing and designing porous materials to enhance the evaporation process. Improved evaporation efficiency is essential for developing more functional materials. The efficiency of porous materials in evaporation is influenced by both their intrinsic properties and structural configuration [29]. Analyzing the intricate interplay between these factors requires a thorough examination of how microstructural features impact evaporation properties. Recently, machine learning has become a valuable tool for exploring the relationship between the structural information of porous materials and their characteristics. This section examines the latest advancements in this field, focusing on identifying coefficient relationships that link the structural characteristics of porous media to their heat and mass transport properties which will affect the evaporation performance.

Permeability, which measures the ease with which fluids can move through the porous structure, directly impacts the supply of liquid to the evaporation front [30]. Higher permeability facilitates better fluid flow, ensuring a steady supply of liquid to replace what is evaporated, thus sustaining evaporation performance. Researchers studied the relationship between structural features and permeability by using machine learning techniques. For instance, Yasuda et al. [31] developed a machine learning framework to elucidate how the structural characteristics of porous materials impact their properties, specifically targeting permeability prediction. The study involved comprehensive training and evaluation of support vector machine (SVM) models, using a combination of filter and wrapper methods for descriptor selection. Initially, redundant descriptors were eliminated using Kendall's correlation, followed by ranking the remaining descriptors with the Relief algorithm. Forward and backward selection processes were then applied to refine the descriptor subset, optimizing prediction accuracy based on mean absolute error (MAE). PCA was used to visualize descriptor spaces, correlating structural differences with prediction errors. This methodology proved effective in reducing computational costs while maintaining high prediction performance. By identifying optimal descriptors, their approach not only improves material characterization but also enables accurate performance prediction for various applications, offering a robust framework for characterizing and designing porous materials.

Similarly, Wang et al. [32] employed deep learning to explore the impact of heterogeneous structures at the pore scale on mineral dissolution and permeability. In their study, the authors used the Monte Carlo random walk method to generate pore structures with varying degrees of heterogeneity and employed the LBM to simulate fluid flow, solute transport, and mineral dissolution processes, with a particular focus on the roles of the Peclet number (Pe) and Damkohler number (Da). They identified and analyzed six dissolution patterns, finding that the heterogeneity of pore structures significantly affects the dissolution patterns and permeability. To study the correlation between permeability and porosity, the research team applied deep neural networks (DNN), the architecture of this DNN model is shown in **Figure 6(a)**. They constructed training and testing datasets based on LBM simulation results to train the DNN. Through training, the DNN learned to recognize the influence patterns of different dissolution modes, Pe, Da, and structural heterogeneity on porosity and permeability. During the testing phase, the DNN demonstrated high accuracy in predicting 12 additional permeability-porosity curves, with an average absolute percentage error of only 6.89%. This indicates that deep learning methods have great potential in predicting complex changes in permeability in heterogeneous porous media with solid dissolution. Compared to traditional LBM methods, DNNs may offer higher accuracy while reducing computational costs.

Effective thermal conductivity (ETC) influences how efficiently heat is transferred within the porous medium, which is crucial for supplying the latent heat

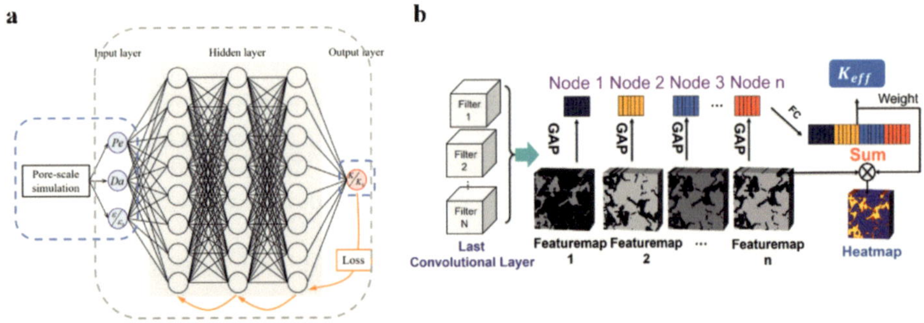

Figure 6.
Models used and Workflows for determining relationship between structural features and properties of porous materials (a) Structure of deep learning model proposed by Wang et al. Reprinted from Ref. [32]. (b) Architecture of model combined by 3D CNN and CAM proposed by Huang et al. Reprinted from Ref. [33].

required for evaporation [34]. High thermal conductivity ensures that heat is uniformly distributed, supporting consistent evaporation rates. Huang et al. [33] adopted an interpretable deep learning strategy combining 3D CNN and Class Activation Mapping (CAM) to do prediction of the ETC of porous structures and visualize the structure-performance relationships. They first used GAN to generate 3D porous material structures with varying porosities and calculated their effective thermal conductivity using PuMA software, thus constructing a dataset. Subsequently, they designed a fully convolutional 3D CNN model without pooling layers, replacing fully connected layers with a Global Average Pooling (GAP) layer to reduce overfitting and maintain spatial information. Using CAM, they generated heat maps to identify key structural regions affecting thermal conductivity. The structure of the model combined by CNN and CAM is represented in **Figure 6(b)**. This method achieved a prediction accuracy of 96.47% on the test set and can iteratively improve material structures through a feedback mechanism, simplifying future work. The performance of the model is represented in **Figure 7(a)**. This study provides new insights and tools for optimizing the design of thermal insulation materials.

Diffusivity, particularly effective diffusivity, dictates the rate at which vapor molecules diffuse through the porous medium [36]. High diffusivity enhances the transport of vapor away from the evaporation front, preventing vapor saturation and maintaining a gradient that drives continuous evaporation. Liu et al. [35] developed

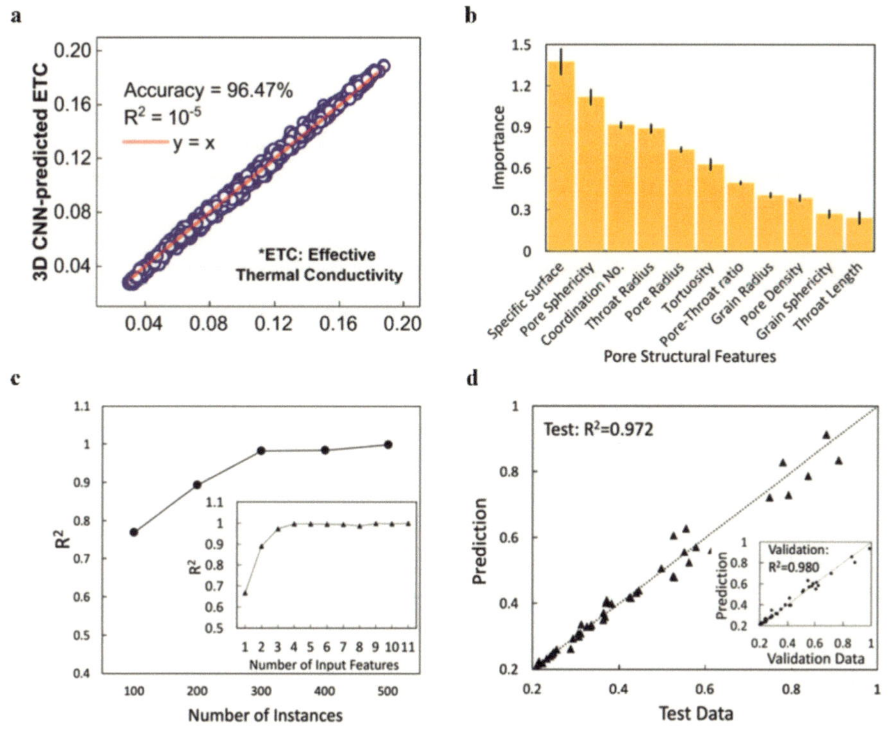

Figure 7.
Performance of model and importance ranking. (a) Performance of the 3D CNNs predictive model on the test dataset. Reprinted from Ref. [33]. (b) Importance ranking of 11 pore structural features estimated by RF model. Reprinted from Ref. [35]. (c) Impact of changing the number of instances and the number of features (inset) on performance of the model. Reprinted from Ref. [35]. (d) The testing and validation performance (inset) of ANN predictions using the three pores structural features from 300 instances. The values represent normalized reaction rates. Reprinted from Ref. [35].

a groundbreaking framework for predicting effective reaction rates using machine learning. This method accurately forecasts reaction rates by analyzing pore structural features, unveiling insightful correlations between structure and reaction rates. They proposed a machine learning model to predict the ETC of porous materials. The process includes extracting structural descriptors, training ML models, and validating their predictions against computed data by using LBM, ultimately providing insights into which structural features most significantly impact the effective thermal conductivity. The results of importance ranking are represented in **Figure 7(b)**. They tested the impact of the number of instances and features on the performance of the model, the result is shown in **Figure 7(c)**. The testing and validation performances of the ANN are represented in **Figure 7(d)**.

The above works investigate the relationship between porous structure characteristics and various heat and mass properties, demonstrating that a similar approach can be used to directly study the relationship between different structural features of porous materials and their evaporation performance. Based on these findings, targeted structural design can be employed to develop porous materials that meet specific usage requirements. However, the current work has some limitations. Firstly, many studies rely on specific types of materials and limited datasets, which may restrict the model's generalization ability and applicability. Future research should focus on collecting and generating more diverse and larger-scale datasets to enhance model robustness. Secondly, although some methods reduce computational costs through feature selection and dimensionality reduction techniques, computational efficiency remains a challenge when dealing with large-scale datasets and high-dimensional feature spaces. Future studies should explore more efficient algorithms and optimization strategies to further reduce computational costs, potentially by combining generative models and deep neural network predictors to create large amounts of data.

4. Prediction of heat and mass transfer fields in porous media

Predicting the mass and heat transfer fields in the evaporation process of porous materials is crucial for optimizing the evaporation performance in various industrial applications and environmental processes. The heat transfer affects the evaporation process by supplying the necessary thermal energy to the liquid within the pores, driving the phase change from liquid to vapor [37]. Similarly, mass transfer governs the movement of vapor out of the porous matrix, influencing the rate and efficiency of evaporation [38]. Accurate predictions of these fields enable the optimization of drying processes, leading to more efficient and uniform evaporation, which is essential for improving the quality of products in industries such as food processing [39], pharmaceuticals [40], and construction [41]. Additionally, optimizing evaporation performance can enhance energy efficiency, reduce processing times, and lower operational costs. Understanding these fields also helps in developing advanced materials with tailored evaporation characteristics, thereby expanding their potential applications and improving overall performance in various industrial contexts.

In contrast, machine learning techniques offer significant advantages by leveraging large datasets to identify patterns and relationships that may not be apparent through traditional methods [42]. Machine learning models can provide more accurate and reliable predictions, adapt to various scenarios with minimal additional data, and significantly reduce the time and cost associated with experimentation [43].

By incorporating machine learning techniques, industries can achieve a deeper understanding of the evaporation processes, leading to enhanced optimization and innovation in product development and process efficiency. In this section, we explore previous works focused on predicting heat and mass transfer properties in porous materials by using machine learning techniques.

4.1 Prediction of heat and mass properties

Previous studies have shown significant progress in predicting transport properties of porous materials using CNNs. For instance, Wu et al. [44] proposed a deep learning model using CNN to predict the effective diffusivity of two-dimensional porous media. This model was developed to address the high computational costs of traditional pore-scale simulation methods and the limitations of empirical formulas. The research team used the LBM to simulate and generate images of porous structures along with their corresponding real effective diffusivity datasets, which were then used to train the CNN. The representatives of generated structures are shown in **Figure 8(a)**. The model underwent preprocessing to remove dead-end paths and trap regions, enhancing the accuracy of predictions for materials with low diffusivity. The architecture of the proposed model is presented in **Figure 8(b)**. In testing, the CNN model demonstrated good performance, with a relative error of less than 10% for high-diffusivity materials, and significantly lower computational costs compared to direct LBM simulations. **Figure 9(a)** represents the performance of the proposed model. Although the relative error was higher for materials with diffusivity below 0.1, the model's performance improved substantially by encoding porosity directly into

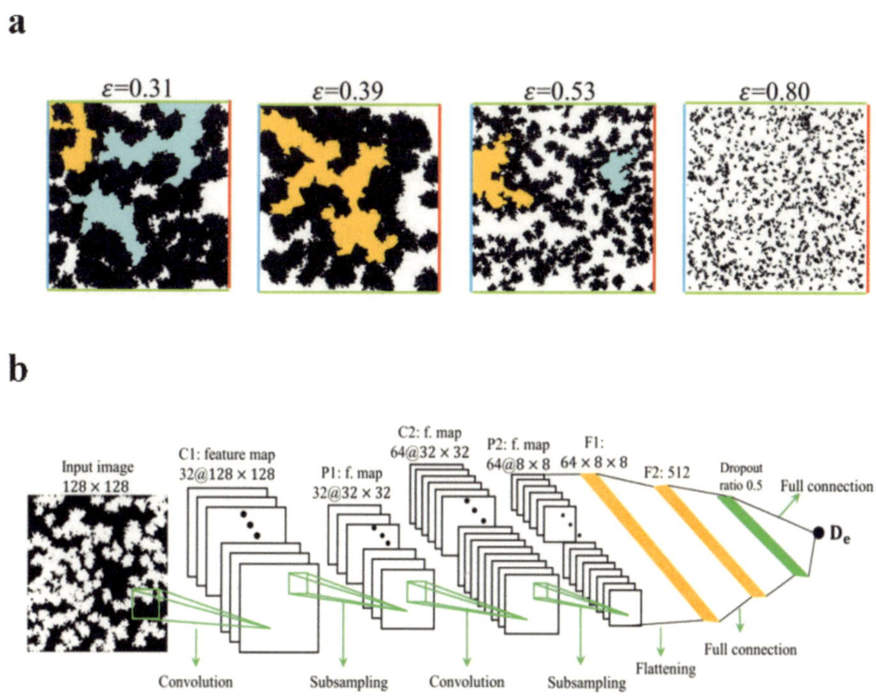

Figure 8.
Architectures of deep learning models (a) Images of reconstructed 2D porous structure for the training. (b) Representation of predicting effective diffusivity by using CNN. Reprinted from Ref. [44].

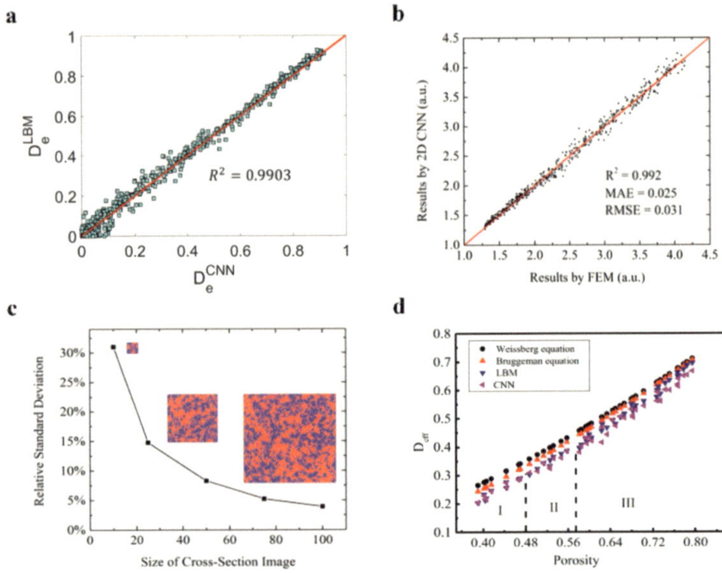

Figure 9.
Prediction results (a) Comparison of the effective diffusivity derived from the LBM and CNN models. Reprinted from Ref. [44]. (b) Comparison between the effective thermal conductivity computed by FEM and 2D CNN model. (c) Relative standard deviation of volume fraction in cross-sectional images with changing image sizes. Reprinted from Ref. [45]. (d) The effective diffusivity estimated by the CNN model and the LBM. Reprinted from Ref. [46].

the neural network and removing physically unfavorable regions, resulting in about 70% of predictions having a relative error of less than 30% for real diffusivity values below 0.1.

Rong et al. [45] proposed a new method for predicting the ETC of composite materials using 2D CNN based on cross-sectional images. The research team designed an optimized 2D CNN model capable of capturing 3D structural information from multiple directional 2D slices of porous structures to predict the ETC of 3D composite materials. They found that using multiple cross-sectional images, especially those along the direction of heat flow, improved the accuracy and representativeness of the predictions. This method demonstrated good generalization capabilities across different types of composite materials (such as composites with spherical particle fillings and random complex structures). A comparison between the ETC computed by the finite element method (FEM) and the 2D CNN model is presented in **Figure 9(b)**. Additionally, the 2D CNN outperformed the 3D CNN in terms of training speed, despite the additional work required to select appropriate cross-sectional images. The representativeness of the 2D cross-section images will affect the performance of the 2D CNN model. If the cross-section images fail to keep the necessary spatial or structural information which is related to the thermal transport properties of the original structure, the prediction accuracy will be reduced. The researchers utilized the volume fraction as a metric for evaluating the representativeness of the cross-section images. The relative standard deviation of volume fraction for cross-section images with varying sizes is represented in **Figure 9(c)**. The innovation of this work lies in using 2D images for predictions, reducing computational costs. However, it also raises several follow-up questions, such as how to select the most representative cross-sectional images and whether this method can be applied to predict other properties that rely on 3D structures. This approach highlighted the potential of using 2D data to predict 3D properties.

Our team [46] proposed a deep learning method based on 3D CNN for predicting the effective diffusivity of porous media, a crucial transport property essential for understanding gas diffusion in porous materials. The specific steps are as follows: create 3D porous structure models using random reconstruction, apply LBM to simulate the effective diffusivity of these structures and establish a training dataset, use the sample structure information self-amplification technique to generate diverse training samples by sliding and selecting existing structures, input the 3D geometric structures and corresponding effective diffusivity into the CNN model for training, and use the trained CNN model to predict effective diffusivity across different porosity ranges.

The results indicate that the proposed model predicts the effective diffusivity with a relative error between 0.026 and 8.95% for porosities ranging from 0.39 to 0.79, significantly outperforming the LBM method while greatly reducing computation time. The effective diffusivity predicted by the CNN model and the lattice Boltzmann method are shown in **Figure 9(d)**. Specifically, for a porosity prediction of 0.5, the CNN required only 3×10^{-4} hours compared to the LBM's 16.96 hours.

4.2 Prediction of heat and mass transfer fields for 3D porous materials

In addition to directly using CNNs to predict thermal and mass properties (such as effective thermal conductivity and diffusivity), CNNs can also be used to predict transfer fields due to their ability to perform image-to-image transformations [47]. This method not only improves the accuracy of predictions but also significantly simplifies the simulation process of complex transfer fields.

Wang et al. [48] proposed a deep learning-based three-dimensional steady-state heat conduction solver to address the challenges of high computational costs and large memory requirements faced by traditional numerical methods like the FEM. They employed a U-net structured deep convolutional neural network, trained using data on 3D geometries and thermal physical parameters generated through COMSOL software and MATLAB. This approach maintained accuracy comparable to the FEM while improving computational speed by approximately 280 times. The study also demonstrated that the framework has good generalization capability when handling unseen complex geometries, with an average relative error typically ranging from 3 to 9%, two histograms showing the distribution of errors are represented in **Figure 10(a)**. Future work may involve extending this framework to 3D transient simulations, requiring further efforts in data generation and framework enhancement.

However, in industrial applications, environmental variables are constantly changing, leading to continually shifting boundary conditions for the porous structures that need to be predicted. Therefore, accurately forecasting the field under these variable boundary conditions presents a significant research challenge. By considering the change of boundary conditions, our team [49] proposed a deep learning-based method for predicting the temperature field in three-dimensional porous media, specifically using a CNN. A model combining a 3D U-Net architecture with 3D VGG blocks was constructed, capable of handling complex three-dimensional structures and predicting temperature fields under various boundary conditions. By utilizing training data generated through LBM, the CNN could quickly and accurately predict temperature fields, with relative errors in effective thermal conductivity ranging from 0.7 to 22.8% (spherical-packed) and 3.1 to 16.0% (irregular). A comparison between the temperature field computed by LBM and predicted by the proposed model is represented in **Figure 10(b)**. Compared to traditional experimental measurements,

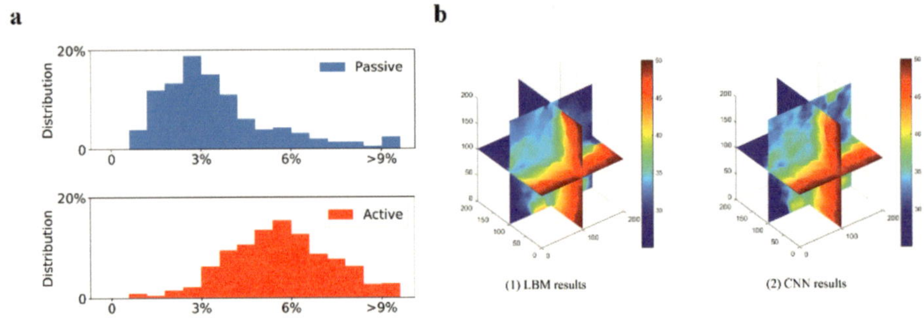

Figure 10.
Prediction results. (a) Statistical analysis of the average relative error on passive and active cases of the open-source dataset. Reprinted from Ref. [48]. (b) Comparison between the temperature field calculated by LBM and CNN model. Reprinted from Ref. [49].

numerical simulations, and empirical models, this method significantly reduces computation time and improves prediction efficiency. However, it requires a large amount of sample data and the model's interpretability is relatively poor. Overall, this work provides an efficient tool for optimizing heat transfer in porous systems.

Ko et al. [50] designed a CNN architecture that includes sub-models trained to learn the spatial relationships of different velocity components. To ensure quality and fluid conservation, the research team implemented a physics-informed loss function, enabling the model to predict complex three-dimensional flow fields rather than relying solely on two-dimensional binary images. The model excelled in predicting flow fields and heat transfer in foam-like structures, demonstrating its accuracy and efficiency, particularly in forecasting permeability and Nusselt numbers. Compared to traditional computational fluid dynamics (CFD) methods, the model significantly reduced computation time and resource requirements. Although currently focused on foam-like structures, the research suggests the potential adaptability of the model to other porous media flow field predictions. The study also discusses optimizing the model to improve prediction accuracy in low-velocity regions. Furthermore, the application of the physics-informed loss function is considered potentially extendable to other physical conservation laws.

In conclusion, these works provide an efficient and accurate tool for predicting heat and mass transfer fields in porous media, reducing the reliance on large-scale experiments and simulations. Future research should focus on collecting and generating more diverse and larger-scale datasets to enhance model robustness. Additionally, exploring more efficient algorithms, such as incorporating attention blocks, and improving model explainability will be the main tasks.

5. Restructure of porous media with best performance

In the pursuit of enhancing evaporation performance, optimizing the structure of porous media is crucial. There are various ways to do performance optimization.

The first method involves optimization based on the relationship between structural descriptors and evaporation performance. Initially, it is necessary to collect structural descriptors of porous materials (such as pore size distribution, porosity, specific surface area, etc.) and the corresponding evaporation performance data. This data can be obtained through experimental measurements or literature searches. The collected data may contain missing values and outliers, so data cleaning is required. Common cleaning methods include filling in missing values, removing outliers, and standardizing and normalizing the data to ensure that all features are on a similar scale, facilitating subsequent model training. After preparing the data, feature engineering is performed. Useful features are extracted from the raw data, such as the mean, variance, maximum, and minimum of pore size. PCA and other methods are then used for feature selection to reduce dimensionality and improve model training speed and performance. Next, appropriate regression models are chosen for performance prediction. Possible models include linear regression [51], SVR [52], random forest regression [53], and gradient boosting regression (such as XGBoost, LightGBM) [54, 55]. Each model is trained and evaluated using k-fold cross-validation (e.g., k = 5 or 10) to prevent overfitting. The predictive performance of each model is assessed using metrics such as Mean Squared Error (MSE), R-squared (R^2), and Root Mean Squared Error (RMSE), and the best-performing model is selected.

Next, a generative model is built to generate porous structures that meet the structural descriptor conditions. First, a large amount of porous structure data is collected, which can be obtained through simulations or experiments. The data is then cleaned to ensure consistency and integrity, and standardized. VAE or GAN is chosen as the generative model. For VAE, the encoder is used to encode the input structural descriptors into latent variables, and the decoder reconstructs the porous structures from these latent variables. The VAE's loss function includes reconstruction loss and KL divergence to ensure that the generated samples are similar in distribution to the original samples. For GANs, the generator produces porous structures, and the discriminator distinguishes between real and generated structures. In cGAN, structural descriptors are included as conditions in both the generator and discriminator to improve the quality of the generated samples. VAEs are trained by maximizing the likelihood, while GANs are trained through adversarial training between the generator and discriminator, incorporating conditional constraints during cGAN training to enhance the conditional conformity of the generated samples. By following this process, the evaporation performance of porous materials can be systematically optimized, improving their effectiveness in practical applications. This method combines the advantages of performance prediction models and generative models, ensuring that the generated porous structures meet the structural descriptor conditions and exhibit excellent evaporation performance.

The second method is based on performance conditions. To directly optimize the evaporation performance of porous materials, a performance-based conditional generative model is employed. In this approach, evaporation performance is used as a condition for generation, with the generative model directly producing porous structures that meet specific evaporation performance requirements. Initially, a large amount of structural data of porous materials and their corresponding evaporation performance data are collected and preprocessed. Conditional Variational Autoencoders (cVAE) or cGAN are chosen as the generative models. For cGAN, evaporation performance is added as a condition in both the generator and discriminator. Specifically, the generator receives random noise and target evaporation performance conditions as input to generate porous structures, while the discriminator receives porous structures and corresponding evaporation performance conditions to distinguish between generated and real structures. In cVAE, the encoder encodes the input porous structures into latent variables, and the evaporation performance condition is combined in the latent space. The decoder then reconstructs porous structures from these latent variables and evaporation performance conditions. cGANs are chosen for their ability to generate high-quality samples, while cVAEs are selected for their capability to generate diverse samples.

During model training, both cGANs and cVAEs face convergence difficulties, particularly in cGANs where the adversarial training requires a large amount of training data. Therefore, it is essential to ensure sufficient training data covering a wide range of evaporation performance. During training, model parameters are continuously adjusted through maximum likelihood estimation (for cVAE) or adversarial loss (for cGAN) to ensure that the generated porous structures meet the evaporation performance conditions and possess reasonable structural characteristics. To address the convergence difficulties in cGAN training, several strategies can be adopted: adjusting the architecture of the generator and discriminator to better handle porous structure data, using more stable optimization algorithms such as the Adam optimizer, and gradually increasing the training data volume to enhance the model's generalization capability. Upon completing model optimization, experimental validation is required to test the actual evaporation performance of the generated porous structures. Experimental results are fed back into the model to further refine the generative model. By iteratively cycling through the process of "generate-evaluate-optimize-validate," the generative model gradually improves its ability to produce porous structures with optimal evaporation performance.

The third method involves combining generative models with optimization algorithms, such as genetic algorithms [56], building on the first method by introducing multi-objective optimization techniques. Multi-objective optimization not only considers evaporation performance but also other important structural characteristics, such as mechanical strength and stability. Genetic algorithms simulate natural selection and genetic variation processes, iteratively optimizing the generated porous structures to achieve global optimality. The workflow is shown in **Figure 11**.

Based on multi-objective optimization, iterative optimization is performed. After the generative model outputs a series of porous structures, the performance

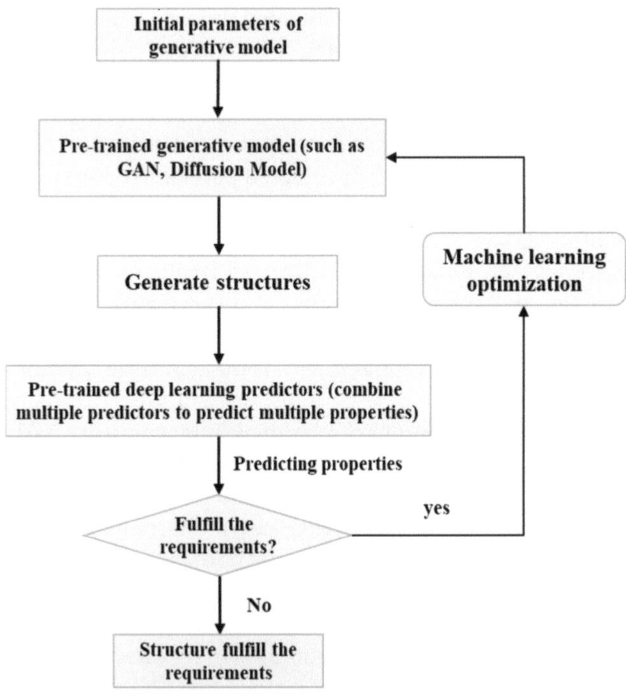

Figure 11.
Workflow of optimization based on the relationship between structural descriptors and evaporation performance.

prediction model is used to evaluate their evaporation performance and other target characteristics. Based on the evaluation results, the input and parameters of the generative model are adjusted, and multiple iterations of optimization are carried out to ensure that the generated porous structures gradually approach optimal performance. In each iteration, evaluation results are continuously fed back, updating the generative model and performance prediction model to ensure that the generated porous structures achieve optimal status within the framework of multi-objective optimization.

6. Conclusions

In this chapter, we explored how machine learning can assist in the intelligent control of evaporation performance in porous media, addressing several key areas. First, the chapter demonstrated the capability of machine learning, particularly deep learning techniques like CNNs and GANs, to generate accurate and representative two-dimensional and three-dimensional porous media structures. These methods surpass traditional approaches in replicating the intricate geometries and material properties of natural porous structures.

Next, we investigated the coefficient relationships between porous media structure and evaporation-related properties (permeability, effective thermal conductivity, and diffusivity). Traditional methods often struggle to fully capture the complexity of these relationships. Machine learning models, however, can analyze extensive datasets to identify patterns and correlations, leading to more efficient and effective material design and optimization.

Following this, the chapter discussed the prediction of heat and mass transfer fields during evaporation in porous media. Traditional simulation methods, while accurate, are computationally intensive. Machine learning offers a powerful alternative by integrating physical principles with advanced algorithms to provide rapid and accurate predictions, thereby enhancing the performance of porous media in practical applications.

Finally, we examined three possible methods for optimizing and generating porous structures with desired properties. These methodologies provide a comprehensive framework for systematically optimizing the evaporation performance of porous materials, ensuring their effectiveness in practical applications. By integrating performance prediction models with generative and optimization algorithms, it is possible to create porous structures that meet stringent performance criteria and contribute significantly to advancements in the field.

Future research should focus on several key areas to further advance the application of machine learning in controlling evaporation performance in porous media. These areas include the diversification of training datasets, the integration of advanced machine learning techniques, improving scalability and automation of processes, fostering interdisciplinary collaboration, and enhancing model interpretability. Addressing these aspects will pave the way for more efficient and effective solutions in various applications involving porous media and evaporation performance.

Acknowledgements

This work was supported by the National Natural Science Foundation of China (No. 52176088).

Conflict of interest

The authors declare no conflict of interest.

Author details

Hui Wang*, Shaoxuan Xu and Zhiguo Qu
MOE Key Laboratory of Thermo-Fluid Science and Engineering, School of Energy and Power Engineering, Xi'an Jiaotong University, Xi'an, Shaanxi, China

*Address all correspondence to: wanghui198@xjtu.edu.cn

References

[1] Lu S, Zhu Q, Ying H. The effect of pore size distribution on the fractal evaporative interface in porous media. Applied Thermal Engineering. 2024;**246**:122960

[2] Obrero JM, Filippin AN, Alcaire M, Sanchez-Valencia JR, Jacob M, Matei C, et al. Supported porous nanostructures developed by plasma processing of metal phthalocyanines and porphyrins. Frontiers in Chemistry. 2020;**8**:520

[3] Koyama M, Suzuki A, Sahnoun R, Tsuboi H, Hatakeyama N, Endou A, et al. Development of porous structure simulator for multi-scale simulation of irregular porous catalysts. Applied Surface Science. 2008;**254**:7774-7776

[4] Siegkas P. A computational geometry generation method for creating 3D printed composites and porous structures. Materials. 2021;**14**:2507

[5] Bose S, Roy M, Bandyopadhyay A. Recent advances in bone tissue engineering scaffolds. Trends in Biotechnology. 2012;**30**:546-554

[6] Bultreys T, Boone MA, Boone MN, De Schryver T, Masschaele B, Van Hoorebeke L, et al. Fast laboratory-based micro-computed tomography for pore-scale research: Illustrative experiments and perspectives on the future. Advances in Water Resources. 2016;**95**:341-351

[7] Fei L, Derome D, Carmeliet J. Pore-scale study on the effect of heterogeneity on evaporation in porous media. Journal of Fluid Mechanics. 2024;**983**:A6

[8] Borgman O, Fantinel P, Lühder W, Goehring L, Holtzman R. Impact of spatially correlated pore-scale heterogeneity on drying porous media. Water Resources Research. 2017;**53**:5645-5658

[9] Fei L, Qin F, Zhao J, Derome D, Carmeliet J. Pore-scale study on convective drying of porous media. Langmuir. 2022;**38**:6023-6035

[10] Paap D, Weinhold B, Chakraborty PP, VandenBos W, Derby MM. Evaporation mechanisms and heat transfer in porous media of mixed wettabilities with a simulated solar flux and forced convection through the media. ASME Journal of Heat and Mass Transfer. 2024;**146**:101602

[11] Kärger J, Valiullin R. Mass transfer in mesoporous materials: The benefit of microscopic diffusion measurement. Chemical Society Reviews. 2013;**42**:4172-4197

[12] Yuan J, Chen W, Tan X, Ma W, Zhou Y, Zhao W. An effective thermal conductivity model of rocks considering variable saturation and pore structure: Theoretical modelling and experimental validations. International Communications in Heat and Mass Transfer. 2021;**121**:105088

[13] Huang J, Xia X, Hu X, Li S, Liu K. A general method for measuring the thermal conductivity of MOF crystals. International Journal of Heat and Mass Transfer. 2019;**138**:11-16

[14] Wang N, Kaur I, Singh P, Li L. Prediction of effective thermal conductivity of porous lattice structures and validation with additively manufactured metal foams. Applied Thermal Engineering. 2021;**187**:116558

[15] Guo Z, Yang J, Tan Z, Tian X, Zhang S, Wang Q. A simplified finite

volume method for effective thermal conductivity in discrete particles. Powder Technology. 2020;**375**:521-532

[16] Wang J, Yuan Q, Dong M, Cai J, Yu L. Experimental investigation of gas mass transport and diffusion coefficients in porous media with nanopores. International Journal of Heat and Mass Transfer. 2017;**115**:566-579

[17] Hong S-N, Yu C-J, Hwang U-S, Kim C-H, Ri B-H. Effect of porosity and temperature on thermal conductivity of jennite: A molecular dynamics study. Materials Chemistry and Physics. 2020;**250**:123146

[18] Zhang X, Yang H, Huang T, Li J, Li P, Wu Q, et al. Research progress of molecular dynamics simulation on the formation-decomposition mechanism and stability of CO_2 hydrate in porous media: A review. Renewable and Sustainable Energy Reviews. 2022;**167**:112820

[19] Yoshino M, Inamuro T. Lattice Boltzmann simulations for flow and heat/mass transfer problems in a three-dimensional porous structure. International Journal for Numerical Methods in Fluids. 2003;**43**:183-198

[20] Wang YD, Meyer Q, Tang K, McClure JE, White RT, Kelly ST, et al. Large-scale physically accurate modelling of real proton exchange membrane fuel cell with deep learning. Nature Communications. 2023;**14**:745

[21] Nguyen PCH, Vlassis NN, Bahmani B, Sun W, Udaykumar HS, Baek SS. Synthesizing controlled microstructures of porous media using generative adversarial networks and reinforcement learning. Scientific Reports. 2022;**12**:9034

[22] Jiang T, Gradus JL, Rosellini AJ. Supervised machine learning: A brief primer. Behavior Therapy. 2020;**51**:675-687

[23] Sagar PL, Maheshbabu C, Teja GVS, Vishnu C, Sai DV, Mahesh C. An evaluation of stock market prediction using supervised machine learning techniques. In: 2023 International Conference on Inventive Computation Technologies (ICICT). 2023. pp. 1-6

[24] Laloy E, Hérault R, Lee J, Jacques D, Linde N. Inversion using a new low-dimensional representation of complex binary geological media based on a deep neural network. Advances in Water Resources. 2017;**110**:387-405

[25] Yang Z, Li X, Catherine Brinson L, Choudhary AN, Chen W, Agrawal A. Microstructural materials design via deep adversarial learning methodology. Journal of Mechanical Design. 2018;**140**:111416

[26] Feng J, Teng Q, He X, Wu X. Accelerating multi-point statistics reconstruction method for porous media via deep learning. Acta Materialia. 2018;**159**:296-308

[27] Shams R, Masihi M, Boozarjomehry RB, Blunt MJ. Coupled generative adversarial and auto-encoder neural networks to reconstruct three-dimensional multi-scale porous media. Journal of Petroleum Science and Engineering. 2020;**186**:106794

[28] Lyu X, Ren X. Microstructure reconstruction of 2D/3D random materials via diffusion-based deep generative models. Scientific Reports. 2024;**14**:5041

[29] Duan H, Fu X, Wang Y. Evaporation properties of water through floating porous structures for desalination. Energy Technology. 2023;**12**:2301126

[30] Kumar N, Arakeri JH. Evaporation from layered porous medium in the presence of infrared heating. Water Resources Research. 2018;**54**:7670-7687

[31] Yasuda T, Ookawara S, Yoshikawa S, Matsumoto H. Machine learning and data-driven characterization framework for porous materials: Permeability prediction and channeling defect detection. Chemical Engineering Journal. 2021;**420**:130069

[32] Wang Z, Chen L, Wei H, Dai Z, Kang Q, Tao W-Q. Pore-scale study of mineral dissolution in heterogeneous structures and deep learning prediction of permeability. Physics of Fluids. 2022;**34**:116609

[33] Huang Q, Hong D, Niu B, Long D, Zhang Y. An interpretable deep learning strategy for effective thermal conductivity prediction of porous materials. International Journal of Heat and Mass Transfer. 2024;**221**:125064

[34] He L, Zhang M, Gu H, Huang A. The influence of thermal radiation on effective thermal conductivity in porous material. International Ceramic Review. 2016;**65**:237-243

[35] Liu M, Kwon B, Kang PK. Machine learning to predict effective reaction rates in 3D porous media from pore structural features. Scientific Reports. 2022;**12**:5486

[36] Wang X, Jia P, Sun S, He X, Lu TJ, Xu F, et al. Evaporation-induced diffusion acceleration in liquid-filled porous materials. ACS Omega. 2021;**6**:21646-21654

[37] Vafai K, Tien HC. A numerical investigation of phase change effects in porous materials. International Journal of Heat and Mass Transfer. 1989;**32**:1261-1277

[38] Talbi M, Prat M. Coupling between internal and external mass transfer during stage-1 evaporation in capillary porous media: Interfacial resistance approach. Physical Review E. 2021;**104**:055102

[39] Onwude DI, Hashim N, Chen G, Putranto A, Udoenoh NR. A fully coupled multiphase model for infrared-convective drying of sweet potato. Journal of the Science of Food and Agriculture. 2021;**101**:398-413

[40] De Meyer L, Lammens J, Mortier STFC, Vanbillemont B, Van Bockstal PJ, Corver J, et al. Modelling the primary drying step for the determination of the optimal dynamic heating pad temperature in a continuous pharmaceutical freeze-drying process for unit doses. International Journal of Pharmaceutics. 2017;**532**:185-193

[41] Van Bockstal P-J, Mortier STFC, De Meyer L, Corver J, Vervaet C, Nopens I, et al. Mechanistic modelling of infrared mediated energy transfer during the primary drying step of a continuous freeze-drying process. European Journal of Pharmaceutics and Biopharmaceutics. 2017;**114**:11-21

[42] Kwon B, Ejaz F, Hwang LK. Machine learning for heat transfer correlations. International Communications in Heat and Mass Transfer. 2020;**116**:104694

[43] Hughes M, Kini G, Garimella S. Status, challenges, and potential for machine learning in understanding and applying heat transfer phenomena. Journal of Heat Transfer. 2021;**143**:120802

[44] Wu H, Fang W-Z, Kang Q, Tao W-Q, Qiao R. Predicting effective diffusivity of porous media from images by deep learning. Scientific Reports. 2019;**9**:20387

[45] Rong Q, Wei H, Huang X, Bao H. Predicting the effective thermal conductivity of composites from cross sections images using deep learning methods. Composites Science and Technology. 2019;**184**:107861

[46] Wang H, Yin Y, Hui XY, Bai JQ, Qu ZG. Prediction of effective diffusivity of porous media using deep learning method based on sample structure information self-amplification. Energy and AI. 2020;**2**:100035

[47] Hou X, Gong Y, Liu B, Sun K, Liu J, Xu B, et al. Learning based image transformation using convolutional neural networks. IEEE Access. 2018;**6**:49779-49792

[48] Wang Y, Zhou J, Ren Q, Li Y, Su D. 3-D steady heat conduction solver via deep learning. IEEE Journal on Multiscale and Multiphysics Computational Techniques. 2021;**6**:100-108

[49] Wang M, Wang H, Yin Y, Rahardja S, Qu Z. Temperature field prediction for various porous media considering variable boundary conditions using deep learning method. International Communications in Heat and Mass Transfer. 2022;**132**:105916

[50] Ko DD, Ji H, Ju YS. Prediction of 3D velocity field of reticulated foams using deep learning for transport analysis. Transport in Porous Media. 2023;**148**:577-604

[51] Su X, Yan X, Tsai C-L. Linear regression. WIREs Computational Statistics. 2012;**4**:275-294

[52] Hearst MA, Dumais ST, Osuna E, Platt J, Scholkopf B. Support vector machines. IEEE Intelligent Systems and their Applications. 1998;**13**:18-28

[53] Breiman L. Random forests. Machine Learning. 2001;**45**:5-32

[54] Dong D, Wen F, Zhang Y, Qiu W. Application of XGboost in electricity consumption prediction. In: 2023 IEEE 3rd International Conference on Electronic Technology, Communication and Information (ICETCI). 2023. pp. 1260-1264

[55] Ke G, Meng Q, Finley T, Wang T, Chen W, Ma W, et al. LightGBM: A highly efficient gradient boosting decision tree. Neural Information Processing Systems. 2017;**30**:52

[56] Katoch S, Chauhan SS, Kumar V. A review on genetic algorithm: Past, present, and future. Multimedia Tools and Applications. 2021;**80**:8091-8126

Chapter 7

Desalinated Seawater Re-Mineralization Using Materials Dissolution

Bin Lu, Jiajiong Xu and Jianbo Lu

Abstract

Desalinated seawater has a strong ability to dissolve solid salt substances (such as pipe scale) and has a low buffering capacity due to the almost complete removal of ions, alkalinity, and hardness. To meet the requirements for protecting the water distribution network and ensuring the health of drinking water, re-mineralization is needed to increase alkalinity and hardness. Based on the principles of convenient operation management and stable and reliable water quality, a re-mineralization dissolution tank is used to increase the alkalinity and hardness of desalinated water. This chapter uses a mathematical model of the re-mineralization dissolution tank to reveal the effects of appropriate limestone-like materials, dissolution tank structure parameters, and operation parameters of the re-mineralization dissolution tank on alkalinity, calcium hardness, dissolved inorganic carbon, pH, and calcium carbonate precipitation potential (CCPP) within the tank. Based on the water quality stability criteria to be met, it was also investigated the cost of consumed limestone-like material and reagent under different operating conditions.

Keywords: desalinated seawater, re-mineralization dissolution tank, water quality stability criteria, LSI, CCPP

1. Introduction

With population growth, urbanization, climate change impacts, and increased water usage in households, agriculture, and industry, the competitive demand for water resources intensifies, and water scarcity is projected to deteriorate further. In the face of freshwater scarcity, seawater desalination is increasingly being utilized to provide potable water [1]. Desalinated seawater serves as a significant water source supplement in China, primarily used for industrial and household purposes, accounting for 66.56 and 33.11%, respectively [2]. Along the Mediterranean coast, desalinated seawater, as a non-traditional water resource, constitutes a primary source of drinking water and is delivered to users through municipal water distribution networks [3, 4]. Additionally, desalinated seawater provides a stable water source for agriculture in numerous counties and cities facing water scarcity [5, 6]. Due to the removal of most ions from the water, desalinated seawater exhibits low hardness, alkalinity, and pH, essentially lacking buffering capacity. During long-term transportation, scaling or corrosion often forms within water distribution pipe networks, achieving a stable reaction balance with the stable hydraulic quality of water transmission. The scale or corrosion surfaces maintain relatively stable morphologies and physicochemical properties [7–9]. Desalinated seawater tends to "dissolve" the scale layers on the inner walls of water distribution pipes, leading to water pollution [10]. As desalinated seawater lacks essential nutrients such as minerals, it also adversely impacts human health and agriculture [11]. Therefore, re-mineralization of desalinated seawater is crucial to increase hardness, alkalinity, buffering capacity, and beneficial mineral ions of desalinated seawater for various end-uses. Re-mineralization of desalinated seawater can typically be achieved through various strategies. Blending with surface water [12, 13], adding chemicals [14], and dissolving minerals [15–17] are common practices. Water quality standards are typically set by national authorities based on an assessment of the trade-offs associated with various uses of desalinated seawater. Lahav et al. proposed a set of quality criteria for the re-mineralization process of desalinated water specified by the Israeli Ministry of Health, considering multiple angles such as chemical stability, biological stability, impact on wastewater treatment, water palatability, health and economic impacts, and post-treatment engineering considerations [18]. Duranceau et al. also suggested water quality targets for post-treated desalinated seawater entering existing water supply systems [19]. The technical guideline, developed and provided by ISO/TC8, addresses the water quality of seawater reverse osmosis desalination products for municipal water supply [20]. Proposed by the Food and Agriculture Organization of the United Nations, a document serves as a field guide for assessing the suitability of water for irrigation [21]. Based on Israeli experiences, Yermiyahu et al. suggested expanding the water quality parameters in seawater desalination facilities supplying water to farmers [22]. The water supply sector is responsible for mapping re-mineralization strategies and selecting appropriate re-mineralization processes. This raises the question of whether there is sufficient technical evidence to justify the efforts and costs associated with various re-mineralization methods, necessitating measures to meet these stringent water quality criteria. This chapter, taking material dissolution as an example, establishes a mathematical model for the re-mineralization dissolution tank based on the thermodynamics and kinetics of material dissolution, represented by a set of differential-algebraic equations. The differential-algebraic equation system is solved using mathematical methods, and evaluated based on the required strict water quality criteria, and technical and economic feasibility.

2. Structure of re-mineralization and dissolution tank

The section of the re-mineralization dissolution tank is shown in **Figure 1**. It operates in an up-flow mode. The re-mineralization packing materials adopt limestone-like filler (such as calcite, dolomite, shell, and so on). The supporting layer uses coarse sand. The flushing method adopts simultaneous air and water flushing, and the flushing cycle depends on the consumption and replenishment of the packing. The flushing process is as follows: first, air flushing for 2 minutes with an air flushing intensity of 60 $m^3/(m^2 \cdot h)$; then simultaneous air and water flushing until the flushed water quality does not exceed the sanitary standard for domestic drinking water, with an air flushing intensity of 60 $m^3/(m^2 \cdot h)$ and a water flushing intensity of 25 $m^3/(m^2 \cdot h)$; finally, single water flushing for 6 minutes with a water flushing intensity of $25 m^3/(m^2 \cdot h)$.

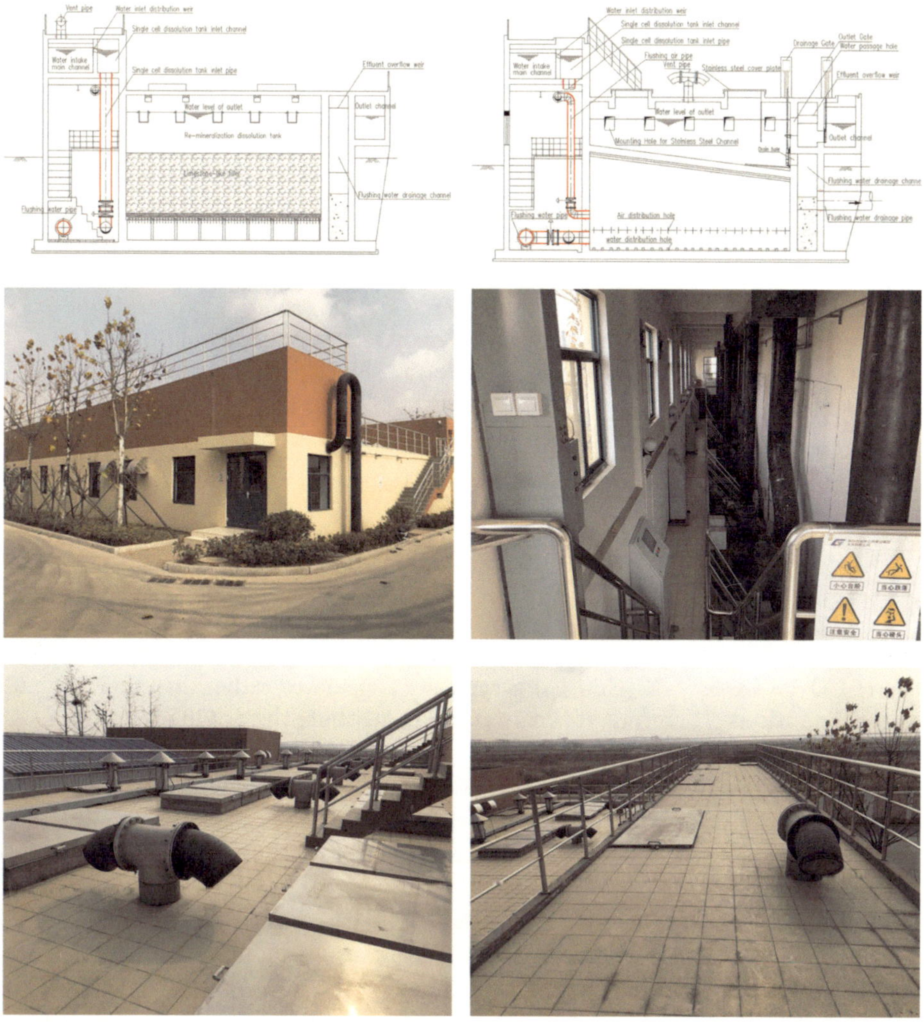

Figure 1.
Section and photos of the re-mineralization dissolution tank in Shandong province of China.

3. Re-mineralization dissolution tank model

The design of re-mineralization dissolution tank refers to the optimal selection of appropriate limestone-like materials, dissolution tank structure parameters, and operation parameters, which affect alkalinity, calcium hardness, dissolved inorganic carbon, pH, and calcium carbonate precipitation potential (CCPP) in the dissolution tank. A re-mineralization dissolution tank model will be established to investigate the effects of appropriate limestone-like materials, dissolution tank structure parameters, and operation parameters of the re-mineralization dissolution tank on alkalinity, calcium hardness, dissolved inorganic carbon, pH, and calcium carbonate precipitation potential (CCPP) within the tank, in order to achieve the optimized design of re-mineralization dissolution tank.

The reaction in limestone-like material dissolution with pre-dosed CO_2 dosing is represented by the following Eq. (1) [23]:

$$CO_2 + H_2O + CaCO_3 \leftrightarrow Ca^{2+} + 2HCO_3 \tag{1}$$

When the desalinated seawater pre-dosed by CO_2 enters the limestone-like filter bed in the re-mineralization dissolution tank, $CaCO_3$ will be dissolved into the bulk solution and the ionic product of Ca^{2+} and $CO_3{}^{2-}$ will be lower than that under the equilibrium condition meeting with the following Eq. (2). It can be recognized that if the dissolution reaction time is long enough, the ionic product of Ca^{2+} and $CO_3{}^{2-}$ will be increasingly close to the equilibrium condition. For limestone-like packed-bed designs, the most convenient formula was proposed by Yamaguchi et al. [24]. According to Yamaguchi et al., the driving force in the dissolution process is the excess concentration of CO_2, namely, the difference between the actual bulk-phase CO_2 concentration and the equilibrium CO_2 concentration. The equilibrium CO_2 concentration is the concentration under equilibrium conditions satisfying Eq. (2) below, where $C_{CO_3^{2-}_eq}$and $C_{Ca^{2+}\_eq}$ are the equilibrium concentration of CO_3^{2-} and Ca^{2+} under the equilibrium condition, Ksp_{CaCO_3} is the ionic product constant of CO_3^{2-} and Ca^{2+}:

$$C_{CO_3^{2-}\_eq} \times \frac{C_{Ca^{2+}\_eq}}{100} = Ksp_{CaCO_3} \tag{2}$$

Based on their respective equilibrium constants, $CO_{2(aq)}$ and $H_2CO_{3(aq)}$ can release protons into water. For simplicity, the sum of the concentrations of these two species (i.e., $H_2CO_3{}^* = CO_{2(aq)} + H_2CO_{3(aq)}$) is typically considered rather than their individual concentrations. Therefore, it is reasonable to attribute the driving force to the difference between the actual bulk $H_2CO_3{}^*$ concentration and the equilibrium $H_2CO_3{}^*$ concentration. The fundamental dissolution rate (r) expression is:

$$r = -\frac{Q \times dC_{H_2CO_3^*}}{dS} = k \times \left(C_{H_2CO_3^*} - C_{H_2CO_3^*\_eq}\right) \tag{3}$$

where Q is the water flow rate, S is the surface area of limestone-like material particles, $C_{H_2CO_3^*}$ is the actual bulk $H_2CO_3{}^*$ concentration, and the $H_2CO_3{}^*$ concentration of $C_{H_2CO_3^*\_eq}$ adheres to Eq. (2) above under equilibrium conditions, where k represents the reaction rate constant. The difference in $Ca(HCO_3)_2$ concentration, derived from Eqs. (1) and (3), can be expressed by the following Eq. (4):

$$dC_{Ca(HCO_3)_2} = -dC_{H_2CO_3^*} = k \times \frac{6 \times (1-\varepsilon)}{d_p \times \varnothing} \times \frac{dH}{u} \times \left(C_{H_2CO_3^*} - C_{H_2CO_3^*{}_{eq}}\right) \tag{4}$$

wherein, ε represents the porosity of the bed, d_p represents the particle diameter, $\varnothing$ represents the particle shape factor (sphere is 1, irregular shapes are less than 1), u represents the superficial velocity, and H represents the height of the packed bed.

Alkalinity (Alk) can be expressed by Eq. (5) below. The equilibrium constant $K1$ (mol/L) for $H_2CO_3{}^*/HCO_3{}^-$ can be represented by Eq. (6) below. The equilibrium constant $K2$ (mol/L) for $HCO_3{}^-/CO_3{}^{2-}$ can be represented by Eq. (7) below. cH is the concentration of hydrogen ion. Kw is the ionic product constant of water:

$$Alk = C_{HCO_3^-} + 2 \times C_{CO_3^{2-}} + 1000 \times \frac{Kw}{cH} - 1000 \times cH \tag{5}$$

$$K1 = \frac{C_{HCO_3^-} \times cH}{C_{H_2CO_3^*}} \tag{6}$$

$$K2 = \frac{C_{CO_3^{2-}} \times cH}{C_{HCO_3^-}} \tag{7}$$

By solving the system of equations (Eqs. 5–7), the concentrations of $H_2CO_3{}^*$ and $CO_3{}^{2-}$ can be calculated using the following equations (Eqs. 8 and 10). Under equilibrium conditions consistent with (Eqs. 5–7), the equilibrium concentrations of $H_2CO_3{}^*$ and $CO_3{}^{2-}$ can also be expressed in (Eqs. 9 and 11), respectively. cHeq is the equilibrium concentration of hydrogen ions under the equilibrium condition:

$$C_{H_2CO_3^*} = \frac{cH \times \left(Alk \times cH + 1000 \times cH^2 - 1000 \times Kw\right)}{K1 \times (cH + 2 \times K2)} \tag{8}$$

$$C_{H_2CO_3^*\_eq} = \frac{cHeq \times \left(Alkeq \times cHeq + 1000 \times cHeq^2 - 1000 \times Kw\right)}{K1 \times (cHeq + 2 \times K2)} \tag{9}$$

$$C_{CO_3^{2-}} = \left(\frac{Alk}{1000} + cH - \frac{Kw}{cH}\right) \times \frac{1000 \times K2}{2 \times K2 + cH} \tag{10}$$

$$C_{CO_3^{2-}\_eq} = \left(\frac{Alkeq}{1000} + cHeq - \frac{Kw}{cHeq}\right) \times \frac{1000 \times K2}{2 \times K2 + cHeq} \tag{11}$$

Subsequently, the concentration of $H_2CO_3^*$ decreases from its initial $C_{H_2CO_3^*}$ to the equilibrium $C_{H_2CO_{3\_eq}^*}$, while the concentration of Ca^{2+} increases from its initial $C_{Ca^{2+}}$ to the equilibrium $C_{Ca^{2+}\_eq}$, a process governed by the law of conservation of mass [25], as shown in the following Eq. (12):

$$C_{H_2CO_3^*} + \frac{C_{Ca^{2+}}}{100} = C_{H_2CO_3^*\_eq} + \frac{C_{Ca^{2+}\_eq}}{100} \tag{12}$$

During the dissolution process of $CaCO_3$, calcium and alkalinity or bicarbonate ions change simultaneously at the same equivalent. The equivalent of dissolved calcium must equal the equivalent of dissolved alkalinity, as shown in Eq. (13):

$$2 \times \frac{C_{Ca^{2+}}}{100} - Alk = 2 \times \frac{C_{Ca^{2+}_{eq}}}{100} - Alkeq \tag{13}$$

Combining Eqs. (2), (9), (11)–(13) and solving the simultaneous equations system, the equilibrium $C_{H_2CO_{3\_eq}^*}$ can be expressed as the function of the bulk terms of $C_{H_2CO_3}$, $C_{Ca^{2+}}$, Alk, and cH as Eq. (14):

$$C_{H_2CO_3^*\_eq} = f\left(C_{H_2CO_3^*}, C_{Ca^{2+}}, Alk, cH\right) \tag{14}$$

It can be seen from Eq. (14) that $C_{H_2CO_3^*\_eq}$ is also a function of $C_{H_2CO_3^*}$, so variable separation cannot be directly applied to integrate Eq. (4). The aforementioned limestone-like packed bed model comprises a set of differential-algebraic equations, and this study proposes for the first time to the use of the finite element differential recursive numerical method for solution. The height of the dissolution bed is divided into 1000 sub-elements (i.e., n = 1000 in the vertical direction). The dimension ΔH of each element is calculated as follows:

$$\Delta H = \frac{H}{n} \tag{15}$$

The alkalinity, calcium ion concentration, and dissolved inorganic carbon concentration at the entrance of the dissolution reactor can be represented by Eqs. (16–18) when the first-bed height difference is observed. Herein, $[Alk]_0$, $[C_{Ca^{2+}}]_0$ and $[Ct]_0$ represent the alkalinity, calcium ion concentration, and dissolved inorganic carbon concentration of the desalinated seawater, respectively. $[CO_2]_0$ denotes the CO_2 dosing concentration injected into the desalinated water prior to entering the limestone-like dissolution bed:

$$Alk[[1]] = [Alk]_0 \ (m\text{mol}/L) \tag{16}$$

$$C_{Ca^{2+}}[[1]] = [C_{Ca^{2+}}]_0 \ (mgCaCO_3/L) \tag{17}$$

$$\text{Ct}[[1]] = [\text{Ct}]_0 + [\text{CO}_2]_0 \ (\text{mmol/L}) \tag{18}$$

The hydrogen ion concentration of $c\text{H}[[1]]$ can be obtained by solving Eq. (19) as follows:

$$\begin{aligned} &\text{Ct}[[1]] \times (2 \times \alpha 1 + \alpha 2) \\ &\quad = \text{Ct}[[1]] \times \frac{K1 \times (c\text{H}[[1]] + 2 \times K2)}{c\text{H}[[1]]^2 + c\text{H}[[1]] \times K1 + K1 \times K2} \\ &\quad = Alk[[1]] + 1000 \times c\text{H}[[1]] - 1000 \times \frac{Kw}{c\text{H}[[1]]} \end{aligned} \tag{19}$$

The bulk concentration of $H_2CO_3^*$ at the first-bed height difference can be calculated from Eq. (8), as shown in Eq. (20):

$$C_{H_2CO_3^*}[[1]] = \frac{c\text{H} \times \left(Alk \times c\text{H}[[1]] + 1000 \times c\text{H}[[1]]^2 - 1000 \times Kw\right)}{K1 \times (c\text{H}[[1]] + 2 \times K2)} \tag{20}$$

According to Eq. (4), the concentration difference of $Ca(HCO_3)_2$ at the i-th height difference can be expressed as Eq. (21):

$$\begin{aligned} &\Delta C_{Ca(HCO_3)_2}[[i]] \\ &\quad = k \times \frac{6 \times (1 - \varepsilon)}{d_p \times \varnothing} \times \frac{\Delta H}{u} \\ &\quad \times \left(C_{H_2CO_3^*}[[i]] - f\left(C_{H_2CO_3^*}[[i]], C_{Ca^{2+}}[[i]], Alk[[i]], c\text{H}[[i]]\right)\right) \end{aligned} \tag{21}$$

The alkalinity, calcium ion concentration, dissolved inorganic carbon, and hydrogen ion concentration at the next difference can be expressed below:

$$Alk[[i+1]] = Alk[[i]] + 2 \times \Delta C_{Ca(HCO_3)_2}[[i]] \tag{22}$$

$$C_{Ca^{2+}}[[i+1]] = C_{Ca^{2+}}[[i]] + 100 \times \Delta C_{Ca(HCO_3)_2}[[i]] \tag{23}$$

$$\text{Ct}[[i+1]] = \text{Ct}[[i]] + \Delta C_{Ca(HCO_3)_2}[[i]] \tag{24}$$

$cH[[i+1]]$ can be obtained by solving Eq. (25) as follows:

$$\begin{aligned} Ct[[i+1]] \times (2 \times \alpha 1 + \alpha 2) \\ &= Ct[[i+1]] \\ &\times \frac{K1 \times (cH[[i+1]] + 2 \times K2)}{cH[[i+1]]^2 + cH[[i+1]] \times K1 + K1 \times K2} \\ &= Alk[[i+1]] + 1000 \times cH[[i+1]] - 1000 \\ &\times \frac{Kw}{cH[[i+1]]} \end{aligned} \tag{25}$$

The bulk H_2CO_3* concentration and equilibrium H_2CO_3* concentration at the next difference are expressed below:

$$C_{H_2CO_3^*}[[i+1]] = \frac{cH[[i+1]] \times \left(Alk[[i+1]] \times cH[[i+1]] + 1000 \times cH[[i+1]]^2 - 1000 \times Kw\right)}{K1 \times (cH[[i+1]] + 2 \times K2)} \tag{26}$$

$$C_{H_2CO_3^*\_eq}[[i+1]] = f\left(C_{H_2CO_3^*}[[i+1]], C_{Ca^{2+}}[[i+1]], Alk[[i+1]], cH[[i+1]]\right) \tag{27}$$

The alkalinity, calcium ion concentration, and dissolved inorganic carbon concentration at the outlet of the limestone-like dissolution bed are expressed as $Alk[[n]]$, $C_{Ca^{2+}}[[n]]$, $Ct[[n]]$. The bulk and equilibrium concentrations of H_2CO_3* at the outlet of dissolution reaction bed are determined as $C_{H_2CO_3^*}[[n]]$, $C_{H_2CO_3^*\_eq}[[n]]$.

4. Water quality for re-mineralization of desalinated seawater

The following parameters are typically used to determine the water quality required for re-mineralizing desalinated seawater, with a focus on health and its interaction with water distribution systems, including the concentration of Ca^{2+}, alkalinity (Alk), pH, Langelier saturation index (LSI) and calcium carbonate precipitation potential (CCPP). LSI can be calculated through the method depicted by literature [26]. CCPP can be calculated through the method depicted by literature [27]. After re-mineralization, the water quality of desalinated seawater needs to meet certain water stability standards before entering the municipal public water supply network. Lahav et al. evaluated the water quality stability standards in terms of chemical stability, biological stability, impact on subsequent wastewater treatment, drinkability, health and economic impact, and feasibility of post-treatment engineering, proposing that the seawater-freshwater re-mineralization process should meet the water quality standards: $Alk > 80$ mg$CaCO_3$/L, 80 < calcium hardness <120 mg$CaCO_3$/L, 3 < CCPP <10 mg$CaCO_3$/L, LSI > 0, pH < 8.5, and these standards have been implemented by the Israeli Ministry of Health [18]. Duranceau et al. also recommended water quality targets for desalinated seawater post-treatment to enter the water supply system [19]. On September 29, 2021, the International Organization for Standardization (ISO) published "Marine Technology - Product Water Quality of Seawater Reverse Osmosis (RO) Desalination - Guidelines for Municipal Water Supply" (ISO 23446:2021), which proposes $Alk \geq 50$ mg$CaCO_3$/L, 60 ≤ calcium hardness

$\leq$120 mgCaCO$_3$/L, $-0.5 \leq$ LSI $\leq$ 0.5, 7.5 $\leq pH \leq$ 8.5. Based on the various water quality stability standards mentioned above and combined with practical experience, CCPP and LSI both play an important role in characterizing water quality properties. Clearly, different values for these parameters can be achieved using different re-mineralization methods. So, these water quality criteria can map the strategies of desalinated seawater re-mineralization processes. This study will conduct technical and economic evaluations of the limestone-like mineral dissolution method using the above-mentioned precise model calculations to justify the efforts and costs incurred in meeting the aforementioned water quality standard limits.

5. Simulation result of re-mineralization dissolution tank

The pre-injected CO_2 desalinated seawater was introduced into a dissolution bed reactor in a typical seawater reverse osmosis plant located in Shandong Province, China, for dissolution and re-mineralization. The bed height (H) of the reactor was 2.7 m, and the superficial velocity was 12.75 m/h. The desalinated effluent water represents dissolved inorganic carbon of 0.062 mmol·L^{-1}, alkalinity of 2.49 mg CaCO$_3$·L^{-1}, calcium ion concentration of 1.15 mg CaCO$_3$·L^{-1} and pH of 6.91. Effects of CO_2 dosage concentration and packing height on the aqueous concentration and equilibrium concentration of $H_2CO_3^*$ are shown in **Figure 2**. It can be seen from the Figure that the aqueous concentration of $H_2CO_3^*$ decreases with increasing the packing height and increases with increasing the CO_2 dosage concentration. The former is attributed to the more consumption of $H_2CO_3^*$ with increasing the residence time in the limestone-like material packing bed. The equilibrium concentration of $H_2CO_3^*$ slowly increases with increasing the packing height increases and significantly increases with increasing CO_2 dosing concentration. These are attributed to the increase of dissolved inorganic carbon resulting to shift carbonate system toward $H_2CO_3^*$.

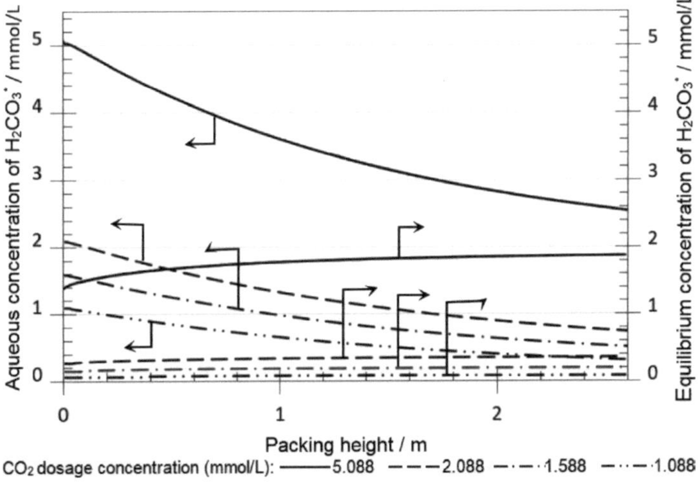

Figure 2.
Effects of CO_2 dosage concentration and packing height on the aqueous concentration and equilibrium concentration of $H_2CO_3^*$.

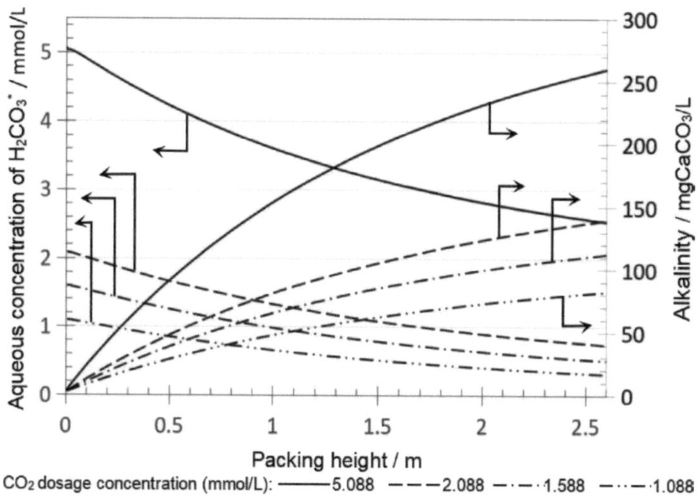

Figure 3.
Effects of CO_2 dosage concentration and packing height on the alkalinity.

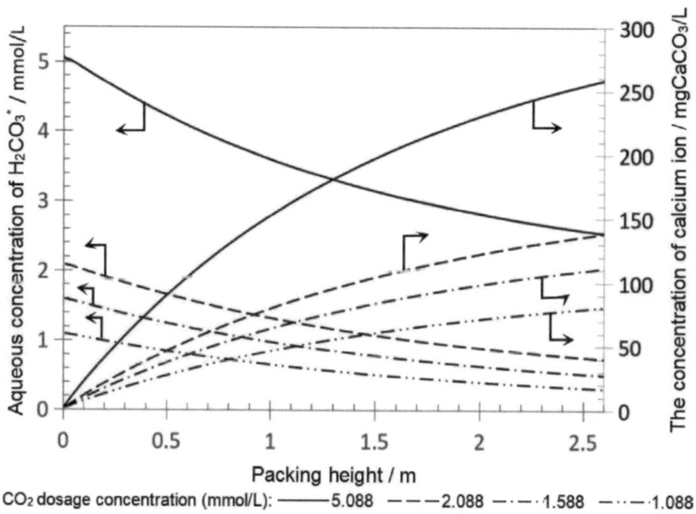

Figure 4.
Effects of CO_2 dosage concentration and packing height on the concentration of calcium ion.

The effects of CO_2 dosage concentration and packing height on the alkalinity are shown in **Figure 3**. As can be seen from **Figure 3**, alkalinity increases with increasing the packing height and CO_2 dosage concentration, respectively. The decrease in the aqueous concentration of $H_2CO_3^*$ simultaneously accompanies the increase in alkalinity. When the CO_2 dosage reached 5.088 mmol/L, 2.088 mmol/L, 1.588 mmol/L, and 1.088 mmol/L, the alkalinity at the outlet of the re-mineralization dissolution tank reached 259.546 $mgCaCO_3/L$, 139.533 $mgCaCO_3/L$, 112.520 $mgCaCO_3/L$, and 82.174 $mgCaCO_3/L$, respectively. According to Eq. (1), when 1 mmol/L of $H_2CO_3^*$ is consumed, 100 $mgCaCO_3/L$ of alkalinity will be produced.

The effect of CO_2 dosage concentration and packing height on the concentration of calcium ions is shown in **Figure 4**. As can be seen from the Figure, the calcium ion concentration increases with increasing the packing height and CO_2 dosage concentration, respectively. The decrease in aqueous concentration of $H_2CO_3^*$ simultaneously accompanies the increase in calcium ion concentration. When the CO_2 dosage reached 5.088 mmol/L, 2.088 mmol/L, 1.588 mmol/L, and 1.088 mmol/L, the calcium ion concentrations at the outlet of the re-mineralization dissolution tank reached 258.206 $mgCaCO_3$/L, 138.193 $mgCaCO_3$/L, 111.18 $mgCaCO_3$/L, and 80.834 $mgCaCO_3$/L, respectively. According to Eq. (1), when 1 mmol/L of $H_2CO_3^*$ is consumed, 100 $mgCaCO_3$/L of calcium ion will be produced.

The effect of CO_2 dosage concentration and packing height on the concentration of dissolved inorganic carbon is shown in **Figure 5**. As can be seen from **Figure 5**, the dissolved inorganic carbon concentration increases with increasing the packing height and CO_2 dosage concentration, respectively. The decrease in aqueous concentration of $H_2CO_3^*$ simultaneously accompanies the increase in dissolved inorganic carbon concentration. When the CO_2 dosage reached 5.088 mmol/L, 2.088 mmol/L, 1.588 mmol/L, and 1.088 mmol/L, the dissolved inorganic carbon concentrations at the outlet of the re-mineralization dissolution tank reached 7.721 mmol/L, 3.520 mmol/L, 2.750 mmol/L, and 1.947 mmol/L, respectively. According to Eq. (1), when 1 mmol/L of $H_2CO_3^*$ is consumed, 2 mmol/L of dissolved inorganic carbon will be produced.

The effect of CO_2 dosage concentration and packing height on pH is shown in **Figure 6**. As can be seen from the Figure, pH increases with increasing the packing height and decreases with increasing CO_2 dosage concentration. The decrease in aqueous concentration of $H_2CO_3^*$ simultaneously accompanies the increase in pH. When the CO_2 dosage reached 5.088 mmol/L, 2.088 mmol/L, 1.588 mmol/L, and 1.088 mmol/L, the pH at the outlet of the re-mineralization dissolution tank reached 6.582, 6.871, 6.948, and 7.035, respectively. $CO_{2(aq)}$ can release a proton to the water according to its own equilibrium constant, resulting in lowering pH.

The effect of CO_2 dosage concentration and packing height on CCPP is shown in **Figure 7**. As can be seen from the Figure, CCPP increases with increasing the packing height and decreases with increasing CO_2 dosage concentration. The decrease in aqueous concentration of $H_2CO_3^*$ simultaneously accompanies the increase in CCPP. When the CO_2 dosage reached 5.088 mmol/L, 2.088 mmol/L, 1.588 mmol/L,

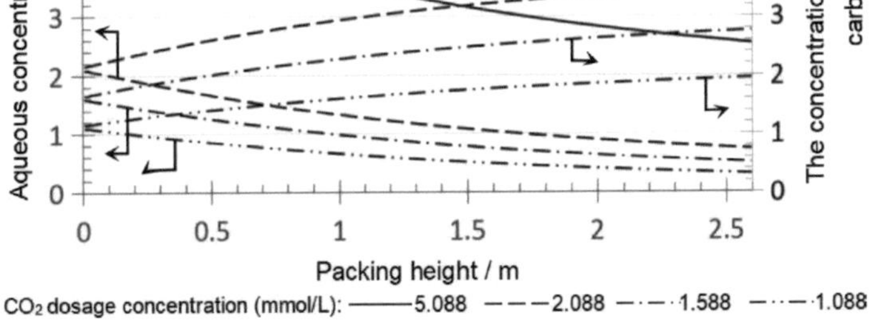

Figure 5.
Effects of CO_2 dosage concentration and packing height on the concentration of dissolved inorganic carbon.

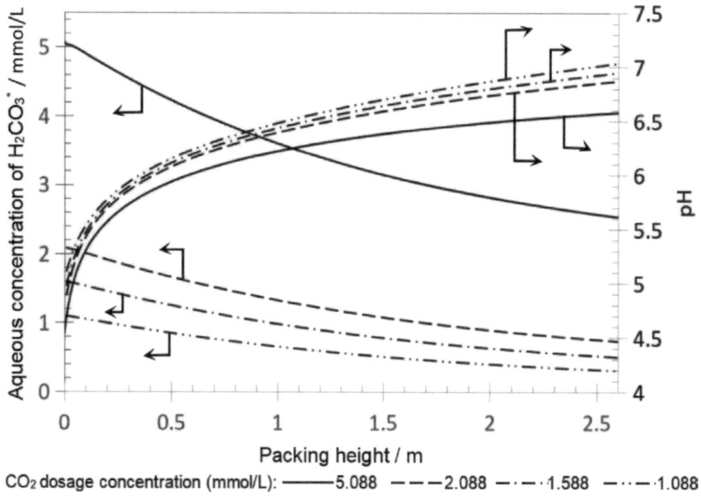

Figure 6.
Effects of CO_2 dosage concentration and packing height on pH.

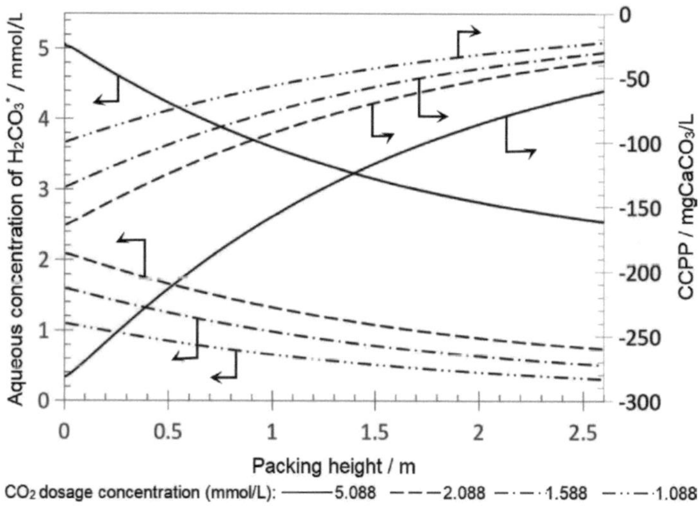

Figure 7.
Effects of CO_2 dosage concentration and packing height on CCPP.

and 1.088 mmol/L, the CCPP at the outlet of the re-mineralization dissolution tank reached −59.786 mgCaCO$_3$/L, −36.422 mgCaCO$_3$/L, −30.066 mgCaCO$_3$/L, and − 22.515 mgCaCO$_3$/L, all of which are negative. This can be attributed to the limited packing height impossible to achieve dissolution equilibrium. The lower *p*H owing to the more CO_2 dosage also results in the lower CCPP. These indicate that the outlet product water from the re-mineralization dissolution tank still has certain corrosion tendency, and *p*H adjustment is still required.

The outlet product water from the re-mineralization dissolution tank is generally added with NaOH to meet the required water quality criteria for CCPP and LSI, aiming for a *p*H increase to approximately 8.3 (i.e., the equivalent point), where the proportion of HCO_3^- is highest in dissolved inorganic carbon, and the alkalinity and

Operating conditions	I	II	III	IV
After re-mineralization using limestone-like material dissolution				
CO_2 dosage/(mmol/L)	5.088	2.088	1.588	1.088
Dissolved $CaCO_3$/(mg$CaCO_3$·L^{-1})	257.056	137.043	110.030	79.684
Dissolved inorganic carbon concentration/(mmol·L^{-1})	7.721	3.520	2.750	1.947
Calcium ion concentration/(mg$CaCO_3$·L^{-1})	258.206	138.193	111.180	80.834
Alkalinity/(mg$CaCO_3$·L^{-1})	259.546	139.533	112.520	82.174
*p*H	6.582	6.871	6.948	7.035
CCPP/(mg$CaCO_3$·L^{-1})	−59.786	−36.422	−30.066	−22.515
LSI	−0.308	−0.530	−0.633	−0.810
After NaOH adjustment				
NaOH dosage for *p*H adjustment/ (mmol·L^{-1})	2.566	0.740	0.507	0.308
Alkalinity/(mg$CaCO_3$·L^{-1})	387.821	176.530	137.863	97.555
Adjusted *p*H with NaOH	8.3	8.3	8.3	8.3
CCPP/(mg$CaCO_3$·L^{-1})	89.096	19.821	11.175	4.410
LSI	1.409	0.900	0.720	0.452
Cost of consumed limestone-like material and reagent/(¥/m^3)	0.942	0.422	0.327	0.229

Table 1.
Regulation of outlet product water from re-mineralization dissolution tank using NaOH adjustment.

acidity are equal [28]. The alkalinity, *p*H, CCPP, and LSI before and after NaOH adjustment are compared, as shown in **Table 1**. As seen from the table, higher CO_2 dosage concentration for desalinated seawater produces higher alkalinity, calcium ion concentration, and LSI for the outlet product from re-mineralization dissolution tank. However, the higher CO_2 dosage concentration is for desalinated seawater, the lower *p*H and CCPP is for outlet product from re-mineralization dissolution tank, requiring the addition of NaOH to adjust the *p*H. If *p*H is adjusted to 8.3 (i.e., equivalent point), the alkalinity, CCPP, and LSI in conditions I to III are all relatively high. For condition IV, the alkalinity is 97.555 mg$CaCO_3$·L^{-1}, calcium ion is 80.834 mg$CaCO_3$·L^{-1}, CCPP is 4.410, and LSI is 0.452. The cost of consumed limestone-like material (taking shell filler as an example) and chemical reagent dosage (used CO_2 and NaOH) for condition IV is approximately 0.229 ¥/m^3.

As can be seen from **Table 1**, when a higher concentration of CO_2 is added to the inlet water of the re-mineralization dissolution tank, a higher alkalinity and calcium hardness can be obtained from the outlet water. So, a portion of the desalinated seawater can bypass the re-mineralization dissolution tank (e.g., with a water flow of $0.75 \times Q$, where Q is the total flow of desalinated seawater), while another portion of the desalinated seawater is added with higher CO_2 dosage (e.g., 5.088 mmol/L) and then injected into the re-mineralization dissolution tank (e.g., with a water flow of $0.25 \times Q$) to dissolve out a higher alkalinity and calcium hardness. After that, the bypassed desalinated seawater is blended with the outlet product water from the re-mineralization dissolution tank and then adjusted with NaOH to regulate *p*H to 8.3. The results are shown in **Table 2**. The alkalinity is 99.057 mg$CaCO_3$·L^{-1}, calcium ion is 65.414 mg$CaCO_3$·L^{-1}, CCPP is 3.540 mg$CaCO_3$·L^{-1}, and LSI is 0.378. The cost of

	Outlet product water from re-mineralization dissolution tank	Bypassing re-mineralization dissolution tank
Water flow	0.25 × Q	0.75 × Q
	After re-mineralization using limestone-like material dissolution	—
CO_2 dosage/(mmol/L)	5.088	—
Dissolved $CaCO_3$/($mgCaCO_3 \cdot L^{-1}$)	257.056	—
Dissolved inorganic carbon concentration/(mmol·L-1)	7.721	0.062
Calcium ion concentration/($mgCaCO_3 \cdot L^{-1}$)	258.206	1.15
Alkalinity/($mgCaCO_3 \cdot L^{-1}$)	259.546	2.49
pH	6.582	—
CCPP/($mgCaCO_3 \cdot L^{-1}$)	−59.786	—
LSI	−0.308	—
Blending desalinated seawater bypassing re-mineralization dissolution tank with outlet product water from re-mineralization dissolution tank		
Dissolved inorganic carbon concentration of blended water/($mmol \cdot L^{-1}$)		1.977
Calcium ion concentration of blended water/($mgCaCO_3 \cdot L^{-1}$)		65.414
Alkalinity of blended water/($mgCaCO_3 \cdot L^{-1}$)		66.754
After NaOH adjustment		
NaOH dosage for pH adjustment/($mmol \cdot L^{-1}$)		0.646
Alkalinity/($mgCaCO_3 \cdot L^{-1}$)		99.057
Adjusted pH with NaOH		8.3
CCPP/($mgCaCO_3 \cdot L^{-1}$)		3.540
LSI		0.378
Cost of consumed limestone-like material and reagent(¥/m^3)		0.236

Table 2.
Regulation of the water blending desalinated seawater bypassing re-mineralization dissolution tank with outlet product water from re-mineralization dissolution tank using NaOH adjustment.

consumed limestone-like material (taking shell filler as an example) and chemical reagent dosage (used CO_2 and NaOH) is approximately 0.236 ¥/m^3. It is worth noting that the construction scale of the re-mineralization dissolution tank can be approximately reduced by 75%.

The re-mineralization dissolution tank model has been established to correlate appropriate limestone-like materials, dissolution tank structure parameters, and operation parameters of the re-mineralization dissolution tank to water quality.

Furthermore, using water quality criteria as constraint conditions, construction and operation costs as the optimization objective function, a multi-factor optimization can be conducted based on controlled variables like characteristics of limestone-like materials, structure parameters of re-mineralization dissolution tank, and operating parameters. As the re-mineralization dissolution tank model is solved in this study using the finite element differential recursion method and the method fails to obtain the analytical solution of the objective function, the gradient descent and backpropagation method may serve as one of the paths to solve the current multi-factor optimization problem.

6. Conclusions

This chapter carried out the optimization of re-mineralization processes (taking minerals dissolution as an example) based on accurate model computation with emphasis on technical efforts and material costs for achieving narrow water quality criteria. A mathematical model of re-mineralization dissolution tank was found based on thermodynamics and kinetics of materials dissolution, and represented a set of differential-algebraic equations, revealing the variations in alkalinity, calcium hardness, dissolved inorganic carbon, pH, and CCPP within the tank as a function of the CO_2 dosage concentration of desalinated seawater and the structural parameters of the tank. Furthermore, the operating costs under different operating conditions were examined based on the required water quality stability criteria. The alkalinity, calcium hardness, and dissolved inorganic carbon increase with the increase in the packing height and with the increase in CO_2 dosage concentration. Conversely, pH and CCPP increase with the increase in the packing height but decrease with the increase in CO_2 dosage concentration. The CCPP of the outlet water from the re-mineralization dissolution tank is negative, indicating a need for pH adjustment. The cost of consumed materials and chemical reagents under operating condition IV is approximately $0.229¥/m^3$.

Author details

Bin Lu[1*], Jiajiong Xu[1] and Jianbo Lu[2]

1 Shanghai Municipal Engineering Design Institute (Group) Co., Ltd., Shanghai, China

2 School of Civil Engineering, Yantai University, Yantai, Shandong, China

*Address all correspondence to: lubin@smedi.com

References

[1] Safe drinking-water from desalination, World Health Organization, 2011

[2] Lin S, Zhao H, Zhu L, et al. Seawater desalination technology and engineering in China: A review. Desalination. 2021; **498**:114728-114741. DOI: 10.1016/j.desal.2020.114728

[3] Sauvet-Goichon B. Ashkelon desalination plant—A successful challenge. Desalination. 2007;**203**:75-81. DOI: 10.1016/j.desal.2006.03.525

[4] Hermony A, Sutzkover-Gutman I, Talmi Y, et al. Palmachim seawater desalination plant—Seven years of expansions with uninterrupted operation together with process improvements. Desalination and Water Treatment. 2015;**55**(9):2526-2535. DOI: 10.1080/19443994.2014.940207

[5] Suwaileh W, Johnson D, Hilal N. Membrane desalination and water re-use for agriculture: State of the art and future outlook. Desalination. 2020;**491**: 114559-114581. DOI: 10.1016/j.desal.2020.114559

[6] Martinez AV, Imbernon MA, Gallego EB, et al. Multidisciplinary assessment of the agricultural supply of desalinated seawater in South-Eastern Spain. Desalination. 2023;**548**: 116252-116267. DOI: 10.1016/j.desal.2022.116252

[7] Zhao L, Liu D, Zhang H, et al. Study on electrochemical reduction mechanisms of iron oxides in pipe scale in drinking water distribution system. Water Research. 2023;**231**: 119597-119605. DOI: 10.1016/j.watres.2023.119597

[8] Sarin P, Snoeyink VL, Bebee J, et al. Physico-chemical characteristics of corrosion scales in old iron pipes. Water Research. 2001;**35**(12):2961-2969. DOI: 10.1016/S0043-1354(00)00591-1

[9] Yang F, Shi B, Gu J, et al. Morphological and physicochemical characteristics of iron corrosion scales formed under different water source histories in a drinking water distribution system. Water Research. 2012;**46**(16): 5423-5433. DOI: 10.1016/j.watres.2012.07.031

[10] Cai M, Zhao Z, Sun W, et al. Impact of pipeline materials on water quality stability of desalinated seawater in the pipeline network. Desalination. 2023; **556**:116558-116569. DOI: 10.1016/j.desal.2023.116558

[11] Lesimple A, Ahmed EF, Hilal N. Remineralization of desalinated water: Methods and environmental impact. Desalination. 2020;**496**:114692-114706. DOI: 10.1016/j.desal.2020.114692

[12] Avni N, Eben-Chaime M, Oron G. Optimization desalinated sea water blending with other sources to meet magnesium requirements for portable and irrigation waters. Water Research. 2013;**47**:2164-2176. DOI: 10.1016/j.watres.2013.01.018

[13] Dreizin Y, Tenne A, Hoffman D. Integrating large scale seawater desalination plants within Israel's water supply system. Desalination. 2008;**220**: 132-149. DOI: 10.1016/j.desal.2007.01.028

[14] Azhar EF, Tahaikt M, Zouhri N, et al. Remineralization of reverse osmosis (RO) desalted water for a Moroccan desalination plant: Optimization and cost evaluation of the lime saturator post. Desalination. 2012;

300:46-50. DOI: 10.1016/j.desal.2012.06.003

[15] Hasson D, Fine L, Sagiv A, et al. Modeling remineralization of desalinated water by micronized calcite dissolution. Environmental Science & Technology. 2017;**51**(21):12481-12488. DOI: 10.1021/acs.est.7b03069

[16] Shemer H, Hasson D, Semiat R. Design considerations of a packed calcite bed for hardening desalinated water. Industrial & Engineering Chemistry Research. 2013;**52**(31):10549-10553. DOI: 10.1021/ie302975b

[17] Shemer H, Semiat R, Hasson D. Re-mineralization of desalinated water using a mixture of CO_2 and H_2SO_4. Desalination. 2019;**467**:170-174. DOI: 10.1016/j.desal.2019.06.017

[18] Lahav O, Birnhack L. Quality criteria for desalinated water following post-treatment. Desalination. 2007;**207**: 286-303. DOI: 10.1016/j.desal.2006.05.022

[19] Duranceau JS, Wilder JR, Douglas S. Guidance and recommendations for posttreatment of desalinated water. Journal—American Water Works Association. 2012;**104**:E510-E520. DOI: 10.5942/jawwa.2012.104.0119

[20] Marine technology—Product water quality of seawater reverse osmosis (RO) desalination—Guidelines for municipal water supply, ISO 23446: 2021

[21] Ayers SR, Westcot WD. Water quality for agriculture. Rome: Food and Agriculture Organization of the United Nations; 1985

[22] Yermiyahu U, Tal A, Ben-Gal A, et al. Rethinking desalinated water quality and agriculture. Science. 2007; **318**:920-921. DOI: 10.1126/science.1146339

[23] Hasson D, Bendrihem O. Modeling remineralization of desalinated water by limestone dissolution. Desalination. 2006;**190**:189-200. DOI: 10.1016/j.desal.2005.09.003

[24] Yamauchi Y, Tanaka K, Hattori K, et al. Remineralization of desalinated water by limestone dissolution filter. Desalination. 1987;**66**:365-383. DOI: 10.1016/0011-9164(87)90218-9

[25] Shemer H, Hasson D, Semiat R. State-of-the-art review on post-treatment technologies. Desalination. 2015;**356**:285-293. DOI: 10.1016/j.desal.2014.09.035

[26] DeMartini FE. Corrosion and the Langelier calcium carbonate saturation index. Journal—American Water Works Association. 1938;**30**(1):85-111. DOI: 10.1002/j.1551-8833.1938.tb19529.x

[27] Rossum JR, Merrill DT. An evaluation of the calcium carbonate saturation indexes. Journal—American Water Works Association. 1983;**75**(2): 95-100. DOI: 10.1002/j.1551-8833.1983.tb05075.x

[28] Birnhack L, Voutchkov N, Lahav O. Fundamental chemistry and engineering aspects of post-treatment processes for desalinated water—A review. Desalination. 2011;**273**:6-22. DOI: 10.1016/j.desal.2010.11.011

Chapter 8

Perspective Chapter: Desalination System Performance Index and Artificial Intelligence

Amr Seoudy

Abstract

Desalination using reverse osmosis theory is a well-established technology to produce fresh water from saline water for more than five decades, the complexity, high cost, and high energy consumption of this process have an effect on plant performance and a negative environmental impact. Performance index for reverse osmosis desalination plants (PIRO) to indicate the performance of the entire desalination system is created, to make it easier for the owners and investors to directly know the reasons for the difference between the planned and actual water cost and correction action to be done. A proposed tool has been created in this study to make the analysis easier for the owners and investors to directly know the reasons for the difference between the planned and actual water cost and the correction action that needs to be done. To lessen the impact of reverse osmosis plants on climate change, an assessment of their environmental impact is required. The environmental performance index for a reverse osmosis desalination plant (ROEPI) has been introduced to indicate its influence on global warming and climate change. Integration between desalination and Artificial Intelligence (AI) allows for increased productivity, reduced energy costs, and minimizes its negative effect on global warming and climate change.

Keywords: reverse osmosis, performance index, climate change, environmental performance index, artificial intelligence

1. Introduction

Desalination using reverse osmosis theory is a well-established technology to produce fresh water from saline water for more than five decades, nowadays it is becoming the most used technology for desalination around the world, as per our synthesis of desalination data suggests that there are 15,906 operational desalination plants producing around 95 million m^3/day of desalinated water for human use, of which 48% is produced in the Middle East and North Africa region [1].

This huge amount of desalinated water produced by different technologies such as multi-stage flash (MSF), multi-effect distillation (MED), electrodialysis (ED), reverse osmosis (RO), nanofiltration (NF), and others technologies, the current production of desalinated water from reverse osmosis now stands at 65.5 million m^3/day, accounting for 69% of the volume of desalinated water produced [1].

With this steady increase in the production of desalinated water using reverse osmosis technology, it has become necessary to find a unified reference to determine the performance efficiency of the plants that operate with the technology of reverse osmosis, and it is important that all stakeholders of the desalination industry conform to this reference, whether they are, manufacturers of plant components, solution providers, and those in charge of operation and maintenance works, owners and investors.

A study has been done to introduce a performance index for reverse osmosis desalination plants (PIRO). The proposed performance index (PIRO) is a calculation of how well the production cost meet its defined cost value along the plant lifetime [2].

Meanwhile, desalination using reverse osmosis has a negative environmental impact that can be generated both in the construction and operation of desalination plants, which causes marine environment pollution, terrestrial environment pollution, and air pollution due to emissions of greenhouse gasses (GHGs) and air pollutants and carbon emission as it is a high energy consumption process. **Figure 1** shows the environmental pollution types in the reverse osmosis system.

A rule must be developed to allow governments, international organizations, owners, and businesses to assess the environmental performance of their plant in an integrated system such as a RO system, which incorporates a large number of equipment and items.

Another study has been done to create an environmental performance index (RO_{EPI}) for reverse osmosis desalination plants. Environmental performance index will help governments, international organizations, and decision-makers assess their facilities and determine the best course of action for minimizing their negative environmental effects [3].

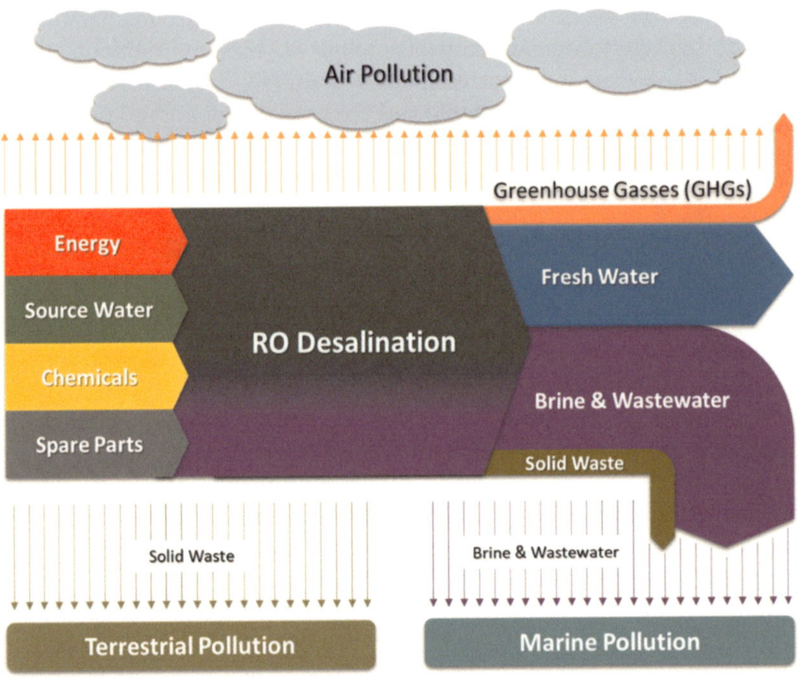

Figure 1.
Environmental pollution types in reverse osmosis system.

1.1 General background

Due to the water demand is continually increasing to meet the needs of population growth, higher living standards, expansion of greenery, increase of per capita consumption, urban development, and industrial expansion, there for new desalination activities and the growing number of large facilities worldwide. Desalination is an energy-intensive process that employs energy to produce freshwater from seawater and rejects the brine back to the sea [4].

Performance index is a calculation of how well work is meeting its defined goal. A long time ago when the reverse osmosis plant started to be wide world used, a lot of studies were done to develop a methods to evaluate reverse osmosis plant performance, most of the developed method considered the membrane's ability to produce fresh water considering quantity and quality in terms of membrane flux and salt rejection, as it is the core of RO unit, after that power and chemical consumption to produce one cubic meter of fresh water is considered as a key factors indicate the performance of the RO system [4].

Desalination plants using reverse osmosis theory normally consist of three main sections, pretreatment, reverse osmosis, post-treatment. Each section contains different items depend mainly on the water source type characteristics and the specifications of the needed water from the treatment process, we consider the water intake as an item in pretreatment section, regardless the type of the intake. Overall performance of entire reverse osmosis desalination plant depends on the performance of each of the plant components [4].

Water desalination has negative environmental impacts on marine environment pollution, air environment pollution, and terrestrial environment pollution. In addition, the desalination plants face several problems, such as corrosion, scale formation, membrane fouling, and sedimentation. The disposal of reject brine from coastal or inland desalination plants and its adverse effects on the ecosystems of the marine environment and groundwater.

The major environmental concerns of desalination are broadly categorized into direct and indirect environmental impacts (EIs). The direct impacts are mainly allied with brine-associated pollution, the emissions of air pollutants such as greenhouse gases (GHGs) emissions and intensified use of energy. The indirect impacts include noise pollution, use of land, and construction related impacts [2].

1.2 Desalination role in aggravate climate change

Desalination process involves removing salt from seawater and filtering it to produce drinking quality water. But the fossil fuels normally used in the energy-intensive desalination process contribute to global warming due to increased levels of greenhouse gases, and the toxic brine it produces pollutes coastal ecosystems.

Desalination is energy-intensive. RO needs up to 6 kWh of electricity per cubic meter of water (depending on both process and its original salt content), though the latest RO plants such as in Perth, Western Australia, and Singapore use 3.5 kWh/m^3, or 4 kWh/m^3 including pumping for distribution. Hence 1 MWe continuous will produce about 4000–6000 m^3 per day from seawater. Nuclear-powered water desalination with RO has a carbon footprint of about 50 gCO2eq/m^3, compared with 1700 g/m^3 for gas and 2900 g/m^3 for coal [5].

However, desalination is an extensive energy process in which to produce 1000 cubic meters of water per day it requires about 10,000 tons of fossil fuel per year [6].

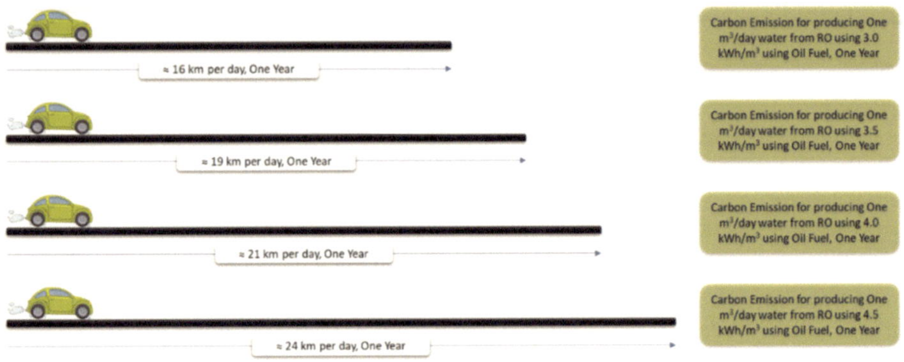

Figure 2.
The carbon intensity of electricity generation varies greatly depending on the fuel source.

Replacing the depleted fossil fuel by renewable and sustainable energy resources becomes a crucial need to decrease the carbon footprint and greenhouse gases emission which they are the main reasons of global warming and climate change.

The World Wide Fund for Nature study found that desalination uses large amounts of energy, emits greenhouse gases, and destroys marine life in some coastal areas. In all cases, desalination affects the environment and climate change, whether this effect is primary or secondary, and vice versa, climate change affects desalination. **Figure 2** shows the carbon intensity of electricity generation varies greatly depending on the fuel source.

2. Research objectives

In last two decades, most of the owners and investors tender documents asking the bidding companies to indicate in their technical solutions not only plant capital expenditure (CAPEX) but also operating expense (OPEX). In last two decades, most of the owners and investors tender documents asking the bidding companies to indicate in their technical solutions not only plant capital expenditure (CAPEX) but also operating expense (OPEX). **Figure 3** shows that in some projects CAPEX is lower than the OPEX and in other projects CAPEX is higher than the OPEX, but the real decision has to be done after calculate freshwater cubic meter total production cost (CAPEX & OPEX).

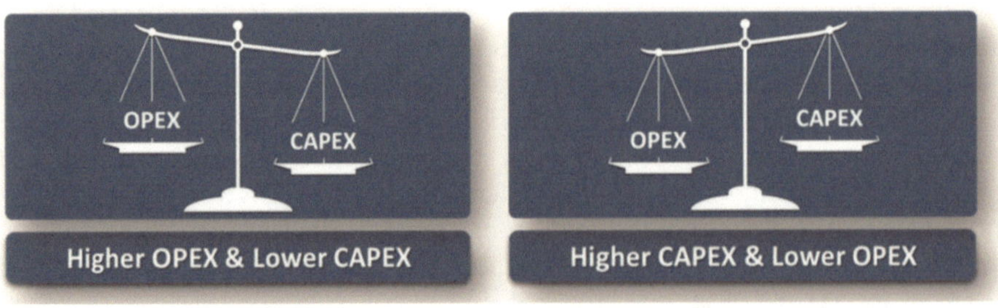

Figure 3.
CAPEX and OPEX cases for RO plant.

Both owners and companies face a challenge to validate or compare design and actual RO total production cost, especially the (OPEX), due to the variation in source water parameters and plant operating condition, either during plant startup or after years of operation.

A rule has to be found to enable both owners and companies to judge their plant performance, in an integrated system such as RO system which contains a lot of equipment and items, each has its performance, it is a big challenge to find an index that can give us a good indication for the entire system [2].

The development of a performance index for reverse osmosis desalination plants (PI_{RO}) to indicate the performance of all the three main sections of the RO desalination plant which include different equipment with it is own performance and with different reference for performance, becomes mandatory.

Meanwhile, the development of the environmental performance index for a reverse osmosis desalination plant (RO_{EPI}) becomes a must to monitor the negative environmental impact of desalination plants and to protect our world from climate change as we can.

The fact that the technical performance, environmental index, and cost are affected by each other and the optimum total performance affects the environmental effect, that is, less performance means higher cost and higher negative environmental effect and vice versa, guides us to consider the comparison between planned and actual production cost for one cubic meter from entire reverse osmosis system to indicate performance index for reverse osmosis desalination plants (PI_{RO}) and to consider the comparison between planned and actual environmental protection cost for one cubic meter from entire RO system to indicate environmental performance index for a reverse osmosis desalination plant (RO_{EPI}).

3. Methodology

3.1 Reverse osmosis desalination plants performance index

Performance index (PI) is the matching of ongoing results against set goals of production cost along the plant lifetime. The proposed performance index (PI_{RO}) is calculated by dividing the actual cost (AC), by the planned cost (PC) for the production of one cubic meter of the entire RO system.

The water production cost of the RO desalination plant for specific design and location is affected by the variation of a lot of parameters, such as feed water quality (i.e., salinity, turbidity, pH, COD, BOD, ... etc.), feed water temperature, membrane aging, chemicals quality, maintenance schedules, shutdown time, and operators' skills.

Two options can be considered for calculation of the planned cost (PC), either according to the design parameters, or according to the startup parameters. The planned cost (PC) according to the design parameters normally considers the variation of feed water temperature and the membrane aging, which mean that we have a different value, which gives us another choice, either consider the lowest value, highest value, the average value or create a value for the planned cost function of time PC (t).

The planned cost (PC) according to the startup parameters consider the actual data for feed water temperature, membrane aging, feed water quality ... etc., at specific time.

As the variation of parameters in the real operation cannot be pre-detected and the change of parameters affect directly the cost, we consider a mirage between the two options as follows:

- Develop a cost graph for the production cost through plant lifetime according to the design parameters PCD (t).
- Develop a cost graph for the production cost through plant lifetime considering the startup parameters and the trend of the graph developed from the design parameters PCS (t), the developed graph will be considered as the reference for evaluation.

A design performance index for the RO system (DPI_{RO}) can be considered to judge and evaluate the plant design. The performance of the entire system is dependent upon the proper design and operation of each component. It is desirable that the RO plant is capable of sustainably producing the required amount of potable water with acceptable quality at minimum cost [4].

3.2 Reverse osmosis plants environmental performance index

Environmental performance index (EPI) is the matching of ongoing negative environmental impact (NEI) results against set goals of production cost along the plant lifetime.

The proposed environmental performance index for reverse osmosis plants (ROEPI) is calculated by dividing the considered environmental evaluation parameters actual cost (EEP_{AC}), by the environmental evaluation parameters planned cost (EEP_{PC}) for the production of one cubic meter of the entire reverse osmosis plants [3].

Due to the variation of operation parameters such as saline water temperature and operation conditions, we consider minimum and maximum values for environmental evaluation parameters planned cost is to be considered as the reference of evaluation for the Environmental Performance Index, as shown in **Figure 4**.

Two options can be considered for the calculation of the planned cost (PC):

- According to the design parameters.
- According to the startup parameters.

The planned cost (PC) according to the design parameters normally considers the variation of feed water temperature and the membrane aging, which means that we have a different value, which gives us another choice, either consider the lowest value, highest value, the average value or create a value for the planned cost function of time PC (t).

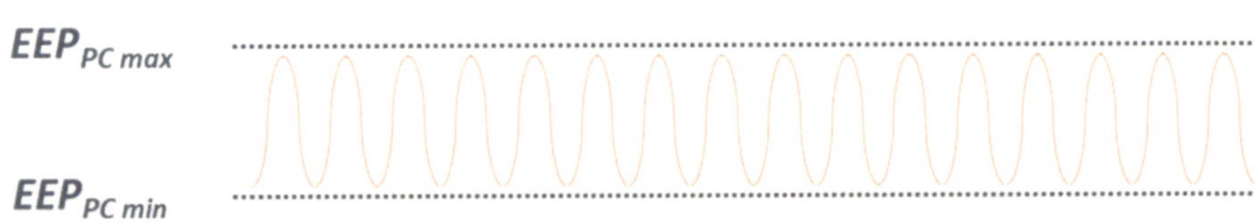

Figure 4.
Maximum and minimum environmental evaluation parameters for RO plant.

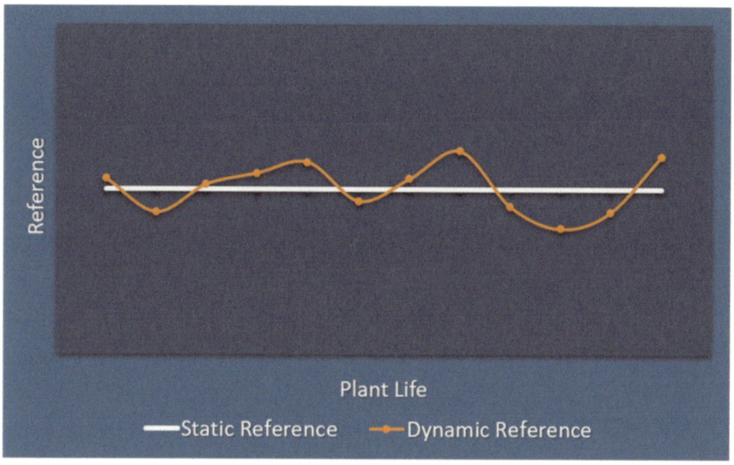

Figure 5.
Environmental performance index dynamic reference.

The planned cost (PC) according to the startup parameters considers the actual data for feed water temperature, membrane aging, feed water quality ... etc., at a specific time.

Due to the variation of governing parameters in RO system, a dynamic reference is considered to be able to evaluate the environmental performance index through the plant lifetime as shown in **Figure 5**.

An environmental performance index based on design for the RO system (RO_{DEPI}) can be consider to judge and evaluate the plant design.

The environmental performance of the entire system is dependent upon the proper design and operation of each component. It is desirable that the RO plant is capable of sustainably producing the required amount of potable water with acceptable quality at a minimum cost.

4. Reverse osmosis performance index calculation

The performance index calculation depends on the calculation of both the actual cost (AC) and the planned cost (PC) for the production of one cubic meter of the entire RO system.

The planned cost or the planned total water cost (PTWC) includes both CAPEX and OPEX as shown in **Figure 6**.

The actual cost or the actual total water cost include both CAPEX and OPEX, the results from calculating the actual total water cost (ATWC) has three probabilities; the first is ATWC less than PTWC, the second is ATWC equal to PTWC, and the third is ATWC more than PTWC. The proposed reverse osmosis performance index for reverse osmosis desalination plants (PI_{RO}) is the deviation of ATWC by PTWC as shown in (Eqs. (1) and (2)).

$$PI_{RO} = \frac{TWC_A}{TWC_P} \tag{1}$$

Where: PI_{RO}, Reverse Osmosis Performance Index; TWC_A, Actual Total Water Cost; TWC_P, Planned Total Water Cost.

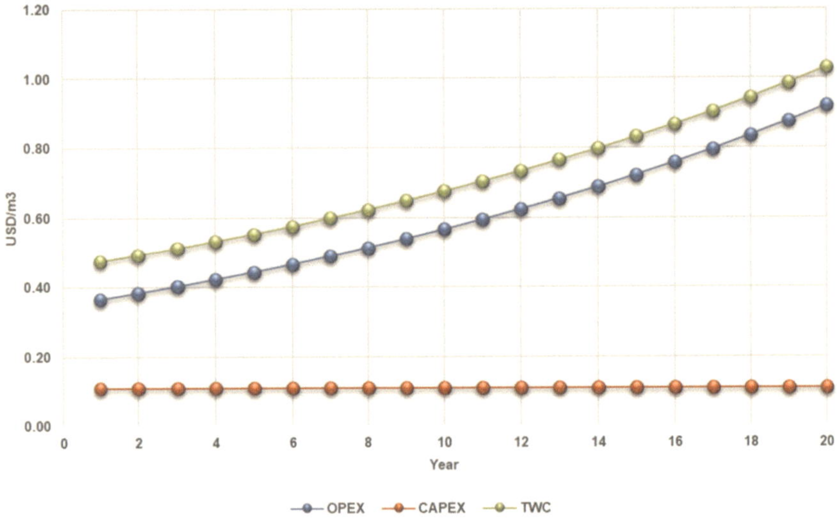

Figure 6.
Planned total water cost include both CAPEX and OPEX.

$$PI_{RO} = \frac{CAPEX_{AC} + OPEX_{AC}}{CAPEX_{PC} + OPEX_{PC}} \tag{2}$$

Where: PI_{RO}, Reverse Osmosis Performance Index; $CAPEX_{AC}$, Capital Expenditure Actual Cost; $OPEX_{AC}$, Operating Expense Actual Cost; $CAPEX_{PC}$, Capital Expenditure Planned Cost; $OPEX_{PC}$, Operating Expense Planned Cost.

The case we have ATWC less than PTWC, seems for the first impression that is the plant performance index is good as the cost of operation is less than expected, maybe this is good indication from investors' point of view, but from a technical point of view, it has two option either that the plant in a good performance and the design is

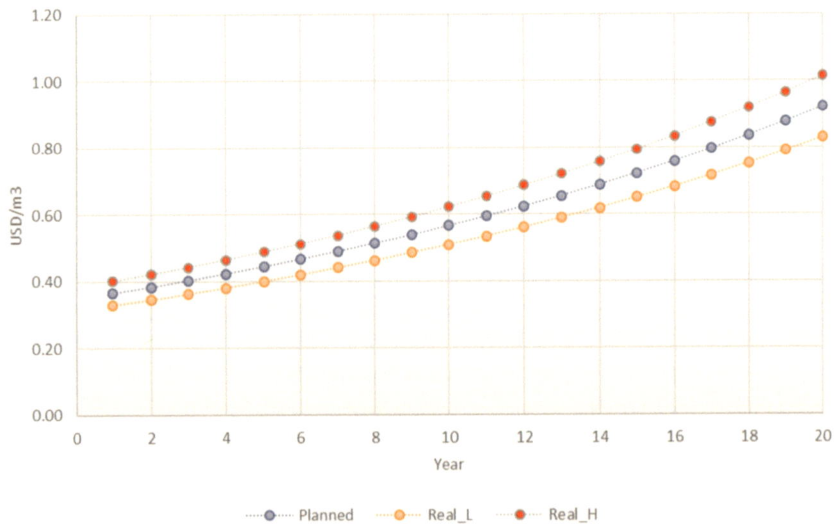

Figure 7.
High and low total water cost for OPEX.

perfect, or maybe the plant not working in the design conditions and this will affect the plant lifetime, so a technical analysis should be done.

The case we have ATWC equal to PTWC, is considered the best scenario as it indicates that everything is going well, but an analysis of all the technical and financial data has to be done. The case where we have ATWC more than PTWC, indicate that something is wrong a deep technical and financial analysis needs to be done.

A certain margin needs to be considered as a low and high value for the actual total water cost as the operation and maintenance of reverse osmosis systems is complicated and contains a lot of items and parameters that affects the calculations, in this study, we will consider this margin only for the OPEX with a value of ±10% of the planned operation cost as shown in **Figure 7**.

5. Performance index calculation tool

A proposed tool has been created in this study to make the analysis easier for the owners and investors to directly know the reasons of the difference between the planned and actual water cost and the correction action needs to be done.

A logic flow chart indicates how we can follow the reason or reasons for the variation depending on checking the difference of the cost of each item which directs us to technical reason if applicable, as example if the tool found that the actual power consumption is higher than the cost of the planned one, it is automatically makes a compare the actual and planned power consumption for all the items previously listed in the tool, then indicate to the operator that as an example the power consumption for the filter feed pump is higher than planned, so a correction action by the operator has to be done to overcome this higher power consumption.

Performance index tool will minimize the time and help the operators and the managers to trace the reasons for the higher operation cost. A part of the logic flow chart is shown in **Figure 8**.

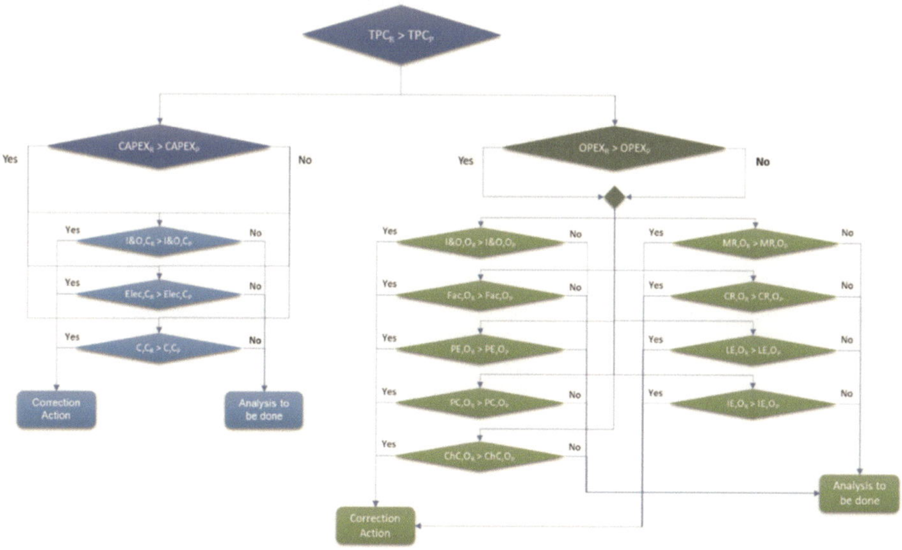

Figure 8.
Logic flow chart for total water cost.

6. Environmental performance index calculation

An evaluation of the environmental impact of reverse osmosis plants become absolutely necessary to reduce it is effect on climate change. The environmental performance index for a reverse osmosis desalination plant (RO_{EPI}) shows the impact of both the design and operation for each plant of global warming and climate change.

A rule has to be found to enable governments, international organizations, owners and companies to judge their plant environmental performance, in an integrated system such as RO system which contains a lot of equipment and items. An environmental performance index for reverse osmosis desalination plants (RO_{EPI}) has been developed to enable the governments, international organization and decision makers to judge their plants and take the correction actions to reducing the environmental impact can be effectively achieved.

The proposed Environmental Performance Index for reverse osmosis desalination plants (RO_{EPI}) consists of three main items as shown in **Figure 9** and (Eqs. (3) and (4)).

$$RO_{EPI} = RO_{MEPI} + RO_{TEPI} + RO_{AEPI} \tag{3}$$

Where: RO_{EPI}, Reverse Osmosis Environmental Performance Index; RO_{MEPI}, Reverse Osmosis Marine Environmental Performance Index; RO_{TEPI}, Reverse Osmosis Terrestrial Environmental Performance Index; RO_{AEPI}, Reverse Osmosis Air Environmental Performance Index.

$$RO_{EPI} = \frac{RO_{EPIAC}}{RO_{EPIPC}} = \frac{RO_{MEPIAC}}{RO_{MEPIPC}} + \frac{RO_{TEPIAC}}{RO_{TEPIPC}} + \frac{RO_{AEPIAC}}{RO_{AEPIPC}} \tag{4}$$

Where: RO_{EPI}, Reverse Osmosis Environmental Performance Index; RO_{EPIAC}, Reverse Osmosis Environmental Performance Index Evaluation Parameters Actual Cost; RO_{EPIPC}, Reverse Osmosis Environmental Performance Index Evaluation Parameters Planed Cost; RO_{MEPIAC}, Reverse Osmosis Marine Environmental Performance Index Evaluation Parameters Actual Cost; RO_{MEPIPC}, Reverse Osmosis Marine

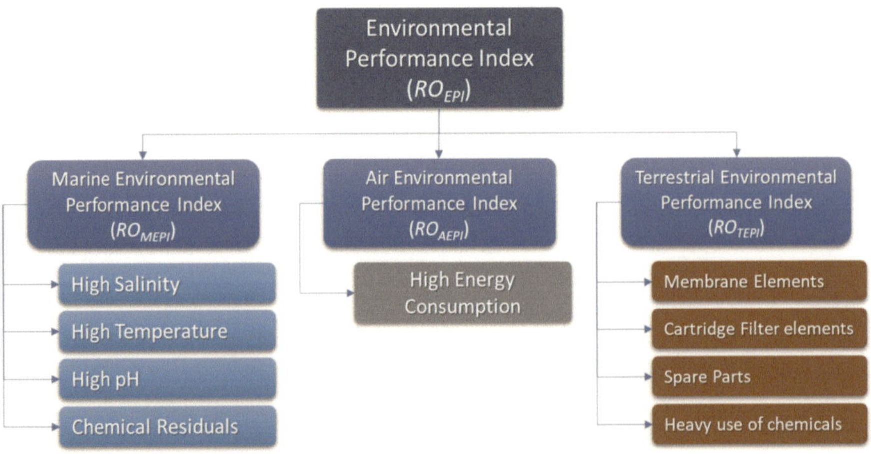

Figure 9.
Reverse osmosis environmental performance index main items.

Environmental Performance Index Evaluation Parameters Planned Cost; RO_{TEPIAC}, Reverse Osmosis Terrestrial Environmental Performance Index Evaluation Parameters Actual Cost; RO_{TEPIPC}, Reverse Osmosis Terrestrial Environmental Performance Index Evaluation Parameters Planned Cost; RO_{AEPIAC}, Reverse Osmosis Air Environmental Performance Index Evaluation Parameters Actual Cost; RO_{AEPIPC}, Reverse Osmosis Air Environmental Performance Index Evaluation Parameters Planned Cost.

Marine environment pollution index (RO_{MEPI}), considering high salinity, high temperature, high pH, and chemical residuals.

The governing equations for the marine environment pollution index (RO_{MEPI}) are shown in (Eq. (4))

$$RO_{MEPI} = RO_{SaEPI} + RO_{TeEPI} + RO_{pHEPI} + RO_{CrEPI} \tag{5}$$

Where: RO_{MEPI}, Reverse Osmosis Marine Environment Pollution Index; RO_{SaEPI}, Reverse Osmosis Salinity Environment Pollution Index; RO_{TeEPI}, Reverse Osmosis Temperature Environment Pollution Index; RO_{pHEPI}, Reverse Osmosis pH Environment Pollution Index; RO_{CrEPI}, Reverse Osmosis Chemical Residuals Environment Pollution Index.

$$RO_{SaEPI} = \frac{Sa_{AC}}{Sa_{PC}} \tag{6}$$

Where: RO_{SaEPI}, Reverse Osmosis Salinity Environment Pollution Index; Sa_{AC}, Salinity Environment Evaluation Parameters Actual Cost; Sa_{PC}, Salinity Environment Evaluation Parameters Planned Cost.

$$RO_{TeEPI} = \frac{Te_{AC}}{Te_{PC}} \tag{7}$$

Where: RO_{TeEPI}, Reverse Osmosis Temperature Environment Pollution Index; Te_{AC}, Temperature Environment Evaluation Parameters Actual Cost; Te_{PC}, Temperature Environment Evaluation Parameters Planned Cost.

$$RO_{pHEPI} = \frac{pH_{AC}}{pH_{PC}} \tag{8}$$

Where: RO_{pHEPI}, Reverse Osmosis pH Environment Pollution Index; pH_{AC}, pH Environment Evaluation Parameters Actual Cost; pH_{PC}, pH Environment Evaluation Parameters Planned Cost.

$$RO_{CrEPI} = \frac{Cr_{AC}}{Cr_{PC}} \tag{9}$$

Where: RO_{CrEPI}, Reverse Osmosis Chemical Residuals Environment Pollution Index; Cr_{AC}, Chemical Residuals Environment Evaluation Parameters Actual Cost; Cr_{PC}, Chemical Residuals Environment Evaluation Parameters Planned Cost.

$$RO_{MEPI} = \frac{Sa_{AC}}{Sa_{PC}} + \frac{Te_{AC}}{Te_{PC}} + \frac{pH_{AC}}{pH_{PC}} + \frac{Cr_{AC}}{Cr_{PC}} \tag{10}$$

Where: RO_{MEPI}, Reverse Osmosis Marine Environment Pollution Index; Sa_{AC}, Salinity Environment Evaluation Parameters Actual Cost; Sa_{PC}, Salinity Environment Evaluation Parameters Planned Cost; Te_{AC}, Temperature Environment Evaluation Parameters Actual Cost; Te_{PC}, Temperature Environment Evaluation Parameters Planned Cost; pH_{AC}, pH Evaluation Environment Parameters Actual Cost; pH_{PC}, pH Evaluation Environment Parameters Planed Cost; Cr_{AC}, Chemical Residuals Environment Evaluation Parameters Actual Cost; Cr_{PC}, Chemical Residuals Environment Evaluation Parameters Planned Cost.

Terrestrial environment pollution index (ROTEPI), considering membrane elements, cartridge filter elements, spare parts, and heavy use of chemicals.

The governing equations for terrestrial environment pollution index (ROTEPI) are shown in (Eq. (11)).

$$RO_{TEPI} = RO_{MeEPI} + RO_{CaEPI} + RO_{SpEPI} + RO_{ChEPI} \tag{11}$$

Where: RO_{TEPI}, Reverse Osmosis Terrestrial Environment Pollution Index; RO_{MeEPI}, Reverse Osmosis Membrane Elements Environment Pollution Index; RO_{CaEPI}, Reverse Osmosis Cartridge Filter Elements Environment Pollution Index; RO_{SpEPI}, Reverse Osmosis Spare Parts Environment Pollution Index; RO_{ChEPI}, Reverse Osmosis Chemical Environment Pollution Index.

$$RO_{MeEPI} = \frac{Me_{AC}}{Me_{PC}} \tag{12}$$

Where: RO_{MeEPI}, Reverse Osmosis Membrane Elements Environment Pollution Index; Me_{AC}, Membrane Environment Evaluation Parameters Actual Cost; Me_{PC}, Membrane Environment Evaluation Parameters Planned Cost.

$$RO_{CaEPI} = \frac{Ca_{AC}}{Ca_{PC}} \tag{13}$$

Where: RO_{CaEPI}, Reverse Osmosis Cartridge Filter Elements Environment Pollution Index; Ca_{AC}, Cartridge Filter Elements Environment Evaluation Parameters Actual Cost; Ca_{PC}, Cartridge Filter Elements Environment Evaluation Parameters Planned Cost.

$$RO_{SpEPI} = \frac{Sp_{AC}}{Sp_{PC}} \tag{14}$$

Where: RO_{SpEPI}, Reverse Osmosis Spare Parts Environment Pollution Index; Sp_{AC}, Spare Parts Environment Evaluation Parameters Actual Cost; Sp_{PC}, Spare Parts Environment Evaluation Parameters Planned Cost.

$$RO_{ChEPI} = \frac{Ch_{AC}}{Ch_{PC}} \tag{15}$$

Where: RO_{ChEPI}, Reverse Osmosis Chemical Environment Pollution Index; Ch_{AC}, Chemical Environment Evaluation Parameters Actual Cost; Ch_{PC}, Chemical Environment Evaluation Parameters Planned Cost.

$$RO_{TEPI} = \frac{Me_{AC}}{Me_{PC}} + \frac{Ca_{AC}}{Ca_{PC}} + \frac{Sp_{AC}}{Sp_{PC}} + \frac{Ch_{AC}}{Ch_{PC}} \tag{16}$$

Where: RO_{TEPI}, Reverse Osmosis Terrestrial Environment Pollution Index; Me_{AC}, Membrane Environment Evaluation Parameters Actual Cost; Me_{PC}, Membrane Environment Evaluation Parameters Planed Cost; Ca_{AC}, Cartridge Filter Elements Environment Evaluation Parameters Actual Cost; Ca_{PC}, Cartridge Filter Elements Environment Evaluation Parameters Planned Cost; Sp_{AC}, Spare Parts Environment Evaluation Parameters Actual Cost; Sp_{PC}, Spare Parts Environment Evaluation Parameters Planned Cost; Ch_{AC}, Chemical Environment Evaluation Parameters Actual Cost; Ch_{PC}, Chemical Environment Evaluation Parameters Planned Cost.

Air environment pollution index (ROAEPI), considering high energy consumption.

The governing equations for air environment pollution index (ROAEPI) are shown in (Eq. (17)).

$$RO_{AEPI} = RO_{EnEPI} \tag{17}$$

Where: RO_{MEPI}, Air Environment Pollution Index; RO_{EnEPI}, Energy Environment Pollution Index.

$$RO_{EnEPI} = \frac{Energy_{AC}}{Energy_{PC}} \tag{18}$$

Where: RO_{EnEPI}, Energy Environment Pollution Index; $Energy_{AC}$, Energy Environment Evaluation Parameters Actual Cost; $Energy_{PC}$, Energy Environment Evaluation Parameters Planned Cost.

7. Environmental performance index indicators

An indicator for the reverse osmosis desalination environmental status is proposed, this indicator shows the environmental status of the station based on the eeverse osmosis desalination environmental index (RO_{EPI}).

The indicator has three cases. **Figure 10** shows that environmental performance index indicator has three cases; the first case is (RO_{EPIAC}) < (RO_{EPIPC}), the second case is (RO_{EPIAC}) = (RO_{EPIPC}), and the third case is (RO_{EPIAC}) > (RO_{EPIPC}).

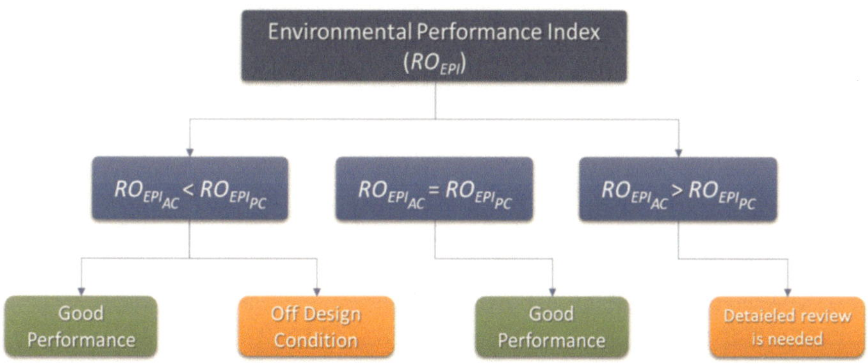

Figure 10.
Environmental performance index indicators cases.

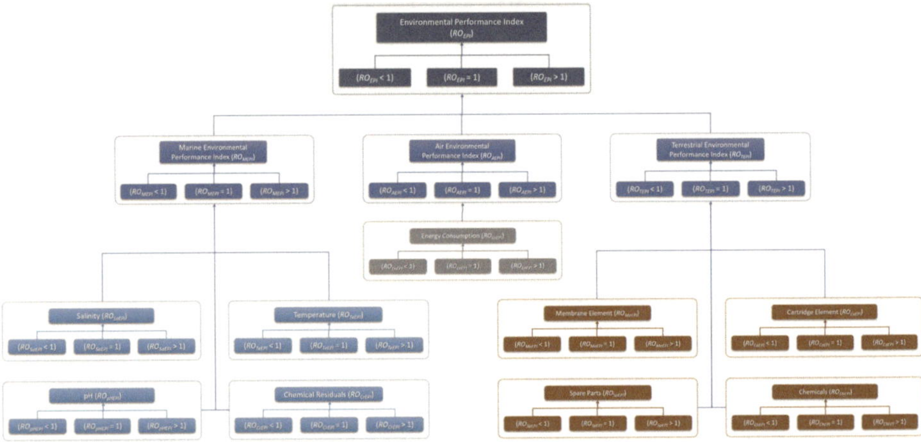

Figure 11.
Detailed environmental performance Index Items Review Sequence.

The first case of (RO_{EPIAC}) less than (RO_{EPIPC}), for first impression this case indicates that the environmental plant performance index is good as the cost is less than expected. From a technical point of view, it has two prospects as follow:

- Either the plant in a good environmental performance and the design is perfect.
- Or maybe the plant not working in the design conditions and this will affect the plant lifetime.

The second case of (RO_{EPIAC}) equal to (RO_{EPIPC}), is considered the best scenario as it indicates that the environmental plant performance index is good and the plant is working in a good manner.

The second case of (RO_{EPIAC}) equal to (RO_{EPIPC}), is considered the best scenario as it indicates that everything is going well, but an analysis of all the technical and financial data has to be done.

The third case of (RO_{EPIAC}) higher than (RO_{EPIPC}), indicates that something is wrong.

A detailed review of all environmental items will identify which prospect is the right one and show which item has a negative effect on the plant. **Figure 11** shows the detailed review sequence. The environmental performance index (RO_{EPI}) will help to indicate the environmental impact of the RO plants and their effect on climate change.

8. Desalination system and artificial intelligence

Desalination systems include water intake, pretreatment (where water is filtered from solid wastes), main desalination stage (salt is removed from water), post-treatment (chlorination and minerals addition), then fresh water is stored in tanks ready to be transported to consumers. Like most manufacturing plants worldwide, desalination plants face multiple challenges that hinder the fluidity of the processes and could pose a threat to their longevity. Employee productivity and safety, machine

DOI: http://dx.doi.org/10.5772/intechopen.1006326

failures, environmental impact, and increased water and energy costs are some examples from these challenges.

The use of artificial intelligence technologies in desalination will improve production efficiency, improve employee productivity, and environmental sustainability.

8.1 Desalination system integration and compatibility

The overall performance of the entire reverse osmosis desalination plant depends on the performance of each of the plant components. A desalination system consists of a lot of equipment, each equipment has its own governing equations and performance evaluation behavior, integration and compatibility of all of these equipment are mandatory. **Figure 12** shows examples of reverse osmosis desalination plant equipment governing equations.

All these processes generate a huge number of interconnected decision parameters/variables throughout the water flow. Analyzing this complex data is often overwhelming, inhibiting the ability of the plant managers and engineers to find adequate solutions in a timely manner.

8.2 Gap between information technology (IT) and operational technology (OT)

All these equipment and instrumentations in the desalination industry have a huge number of data, variables, and parameters affected by the operation parameters, equipment performance and water production target. High-skill operators, engineers, and managers are needed to deal with these daily data and variables. Analyzing this complex data is often overwhelming, inhibiting the ability of the plant managers and engineers to find adequate solutions in a timely manner.

With the right approach and mindset, the industry challenges mentioned earlier can be converted into opportunity areas but first we have to think about them in a more general way. In a broader context, most of these challenges can become a gap between information technology (IT) and operational technology (OT). Due to the lack of common software interfaces, standard data formats, and common connectivity protocols between the two technologies, organizations are struggling with the IT and OT worlds which have largely occupied separate domains [7].

Figure 12.
Reverse osmosis desalination plant equipment governing equations examples.

Machine Learning (ML) which is a subset of artificial intelligence (AI) can analyze these vast production datasets to predict and adapt to certain scenarios and thus convert challenges into solutions, optimization, and improvement with the right approach and mindset, the desalination industry challenges can be converted into opportunity areas but first we have to think about them in a more general way.

These challenges can become a gap between information technology (IT) and operational technology (OT), this gap between the two technologies forces the desalination organizations to struggle with (IT) and (OT) until a common software interface can be implemented.

8.3 Intelligent desalination system

Intelligent desalination plants leverage interconnected IT-OT systems plants taking advantage of Industry 4.0, featuring machines and software capable of self-diagnosis and alerting potential malfunctions. The introduction of advanced machines and devices equipped with smart sensors by IoT facilitates continuous streams of data to be uploaded to the cloud for analysis. This real-time data collection allows for immediate monitoring and assessment of various parameters within the desalination process. Consequently, any potential issues can be detected and addressed promptly, reducing downtime and increasing operational efficiency.

Once the data is collected, it is processed using artificial intelligence (AI) and machine learning (ML) technologies. These systems continuously improve with the availability of more data, enhancing their predictive capabilities and decision-making accuracy. By integrating intelligent connected equipment within water production units, the desalination process benefits from data-driven decision-making. This integration leads to higher efficiency and productivity, as the system can optimize operations based on real-time insights and historical data trends.

IoT sensors play a crucial role in gathering real-world data about water flow, such as temperature, pressure, and chemical composition. They also monitor the health of machinery by tracking temperature, vibrations, and operating conditions. Different pieces of equipment, including pipes, pumps, and valves, are connected through IoT, facilitating comprehensive communication throughout the desalination process. This interconnectedness ensures that all components of the system are functioning harmoniously and that any anomalies are quickly identified and rectified.

While IoT is primarily concerned with the real-time gathering of production data, AI and ML use mathematical algorithms to analyze this data and extract meaningful insights. These insights enable better decision-making and optimization of the desalination process. The high volume, velocity, and variety of IoT data often require big data analytics technologies to manage and interpret the information effectively. By leveraging AI, these technologies can significantly improve manufacturing efficiency and decision-making, leading to more sustainable and cost-effective desalination operations. Artificial intelligence technologies help the industry solve real-world desalination problems and promptly make more practical decisions.

9. Conclusion

Performance index (PI_{RO}) will enable both owners and companies to judge their plant performance, in an integrated system such as an RO system which contains a lot of equipment and items, each has its performance, it is a big challenge to find an index that can give us a good indication for the entire system. Performance index (PI_{RO}) is calculated by dividing the actual cost (AC), by the planned cost (PC) for the production of one cubic meter of the entire RO system. A design performance index for the reverse osmosis system (DPI_{RO}) can be considered to judge and evaluate the plant design. Performance index tool will minimize the time and help the operators and the managers to trace the reasons for the higher operation cost.

Reverse osmosis environmental performance index (RO_{EPI}) will enable both owners and companies to judge their plant environmental performance, in an integrated system such as RO system which contains a lot of equipment and items, each has its environmental performance, it is a big challenge to find an index that can give us a good indication for the entire system environmental performance.

Reverse osmosis environmental performance index (RO_{EPI}) is calculated by dividing the actual cost (AC), by the planned cost (PC) for the considered environmental evaluation parameters. The environmental performance index (RO_{EPI}) will help to indicate the environmental impact of the RO plants and their effect on climate change.

The use of artificial intelligence technologies in desalination will improve production efficiency, improve employee productivity, and environmental sustainability. Artificial intelligence technologies help the industry solve real-world desalination problems and promptly make more practical decisions.

Author details

Amr Seoudy
Water Treatment Consulting Office (WTCO), Cairo, Egypt

*Address all correspondence to: a.seoudy@wtco-eng.com